THE LOCAL HISTORIANS OF ATTICA

AMERICAN PHILOLOGICAL ASSOCIATION

Monograph Series

T. Robert S. Broughton, editor

Number 11

The Local Historians of Attica
by
Lionel Pearson

THE LOCAL HISTORIANS OF ATTICA

Lionel Pearson

Scholars Press
Atlanta, Georgia

A reprint of the 1942 edition published
by The American Philological Association

THE LOCAL HISTORIANS
OF ATTICA

Lionel Pearson

© 1981
The American Philological Association

2nd Printing

Library of Congress Cataloging in Publication Data

Pearson, Lionel Ignatius Cusack
    The local historians of Attica.

    Reprint. Originally published: Philadelphia, Pa. :
    American Philological Association, 1942. (Philological
    monographs ; no. 11) (ISSN 0278-06427)
    Revision of thesis (Ph.D)—Yale, 1939.
    Includes bibliographies.
    1. Greece—Historiography. 2. Historians—Greece.
I. Title. II. Series: Philological monographs ; no.11.
DF211.P4 1981   938'.5'0072   81-16556
ISBN 0-89130-540-8   AACR2

Printed in the United States of America
on acid-free paper

TO MY FATHER
ARTHUR A. PEARSON

PREFACE

Like Greek philosophy after Socrates, Greek historiography in the fourth century B.C. developed along more lines than one. The rhetorical style of history writing was cultivated especially by the pupils of Isocrates and its characteristics are revealed fairly clearly in later authors who tried to follow in their footsteps. Very different from this rhetorical history with its elaborate moralizing were the dry chronicles of local Athenian history written by Cleidemus and Androtion and others, who called their works *Atthides* and apparently took for their model the first writer to compose a specifically Athenian history, Hellanicus of Lesbos. With a strong emphasis on religious history, mythological interpretation, and, where possible, accurate chronology, they established a distinct literary tradition which endured into the following century when Philochorus wrote an *Atthis* that came to be more widely read than any other.

No specimen of an *Atthis* has survived. But the development and continuity of the Atthid tradition is often taken for granted as an established fact of literary history and the qualities which its adherents shared are more generally known than the fragments of individual Atthidographers. When a conventionalized critical opinion has been generally accepted and the reasons for it have been largely forgotten, the time seems ripe for a new presentation and a new study of the evidence. This monograph, therefore, is a study of the characteristics of the local historians in the fourth and third centuries B.C.—the Atthidographers—as revealed in their fragments. It is concerned with their individual peculiarities as well as with the qualities which they had in common and with their loyalty to a literary tradition, the beginnings of which must be sought in early Ionian historiography.

Since the works of these and almost all other Greek historians in the fourth and third centuries have been lost, we are obliged to study the history of Athens during that period from other sources and our knowledge of the development of Greek historical writing between Xenophon and Polybius is dependent on fragmentary material. Fragments—a misleading term to the uninitiated, since papyrus fragments of the original texts play a much smaller part than quotations and criticisms in the works of later Greek authors— are notoriously treacherous as sources for literary history. Authors who indulge freely in quotation are quite capable of misunderstanding and misrepresenting their predecessors and their criticisms are, at best, one-sided. Any verdict, therefore, passed on the work of an author which is not extant, must be provisional. The danger is that criticism of lost works may become conventionalized and that the pattern of literary development may be unduly simplified. When the papyrus texts of Menander and Callimachus were discovered, it became clear that mistakes could be made in attempting reconstruction on the basis of later quotation and comment. But no such object lesson is available for us when we attempt to reconstruct the lost works of Greek historians. The *Hellenica of Oxyrhynchus*, which is by far the most important historical fragment discovered on papyrus, is still a work of disputed authorship and its discovery has not enabled us to revise our views of Ephorus or Theopompus. It might be possible, perhaps, to shield some student carefully from direct contact with the work of Thucydides and notice how far astray he would go in attempting to reconstruct his history after reading Dionysius of Halicarnassus and other ancient critics and laboriously collecting his "fragments" from lexicographers, scholiasts, and anthologists. It might be possible to make such an experiment; but its results would scarcely justify its cruelty, especially since the days of fragment collecting are now almost over.

For myself, indeed, I admit that I am a mere mortal and have never collected fragments of historians. The tremendous work done by the brothers Karl and Theodor Müller, in a day when comprehensive indices were few, is too rarely appreciated; but without the first volume of their *Fragmenta Historicorum Graecorum* the present study would not and could not have been attempted. The fragments of local Attic histories, except for the *Atthis* of Hellanicus, have not yet appeared in Felix Jacoby's *Fragmente der griechischen Historiker*. Professor Jacoby has for some time been busy in Oxford preparing the volume which is to contain them, but the publication of new volumes of his *Fragmente*, like many other scholarly undertakings, has been interrupted by the war. There are a number of new fragments which should be added to the Müller collection and must be taken into account in any discussion of the Atthidographers; and it will be extremely valuable to have them collected in one

place together with any further additions that Professor Jacoby
may have discovered himself. In the meantime, one is dependent
on the standard bibliographies and works of reference for information about the new fragments, from authors previously neglected
or unknown, which have been discovered in the last seventy years.
The places in which these new fragments were published and discussed for the first time are indicated in this book in the footnotes
and in the bibliographies at the end of the sections in Chapters IV
and VI.

Some of the fragments of the Atthidographers are well known
as evidence for controversial points in Athenian history, especially
constitutional history and the problems connected with Solon and
Cleisthenes and the tyranny of Peisistratus. I have not attempted
in many cases to offer new solutions of these problems or even to
decide in each instance between alternative solutions; nor have I
attempted to list all the literature in which these problems are discussed. This does not mean that I have overlooked such problems
or regard them as irrelevant to a study of this kind. The difficulty
is that, when there is uncertainty over historical details, it is
impossible to pass final judgment on the trustworthiness of an
Atthidographer. When that is the case, one can only hope to
clarify the issue by attempting to estimate the general character of
the writer from other evidence. It is in this respect that an investigation into literary history, such as this book offers, may be
of some value to the student of political history.

This study was written in the first place as a dissertation which
was accepted by the Graduate School of Yale University in 1939
in partial fulfilment of the requirements for the degree of Doctor
of Philosophy. Since that time it has been thoroughly revised and
partially rewritten, but the changes have been in matters of detail
and manner of presentation; the conclusions remain essentially
unaltered. It owes a great deal to the careful criticism of Professor
A. M. Harmon, who supervised the writing of the original dissertation and has given me valuable help in the work of revision. Others
who have enjoyed the privilege of his advice and his penetrating
scholarship will understand how happy I am to acknowledge my
indebtedness to him. To Professor M. I. Rostovtzeff and Professor
C. B. Welles I am also deeply obliged for suggestions in matters of
detail as well as for their constant encouragement. Nor must I
forget my colleague at Stanford, Professor Hermann Fränkel, who

has been responsible for some important changes in the final version
of the manuscript. Last of all, I take the opportunity to acknowledge most gratefully the assistance and coöperation given to me
by the Monograph Committee of the American Philological Association and by the present editor of the Association, Professor
T. R. S. Broughton.

LIONEL PEARSON

STANFORD UNIVERSITY,
April 1942.

CONTENTS

LIST OF ABBREVIATIONS

Abbreviations of names and works of Greek and Latin authors follow, with a few exceptions, the systems employed in the new edition of Liddell and Scott, *Greek-English Lexicon*, and the *Thesaurus Linguae Latinae*, respectively; those of titles of periodicals the system of J. Marouzeau in the *Année Philologique*. A list of abbreviations used for collections and periodicals referred to in this monograph is given below.

AAT—Atti della r. Accademia di Scienze di Torino.
AHR—American Historical Review.
AJPh—American Journal of Philology.
CAH—Cambridge Ancient History.
CPh—Classical Philology.
CQ—The Classical Quarterly.
CRAI—Comptes rendus de l'Académie des Inscriptions et Belles-Lettres.
FGrH—Fragmente der griechischen Historiker, ed. F. Jacoby.*
FHG—Fragmenta Historicorum Graecorum, ed. C. and Th. Müller.
H—Hermes.
HSPh—Harvard Studies in Classical Philology.
IG—Inscriptiones Graecae.
JHS—Journal of Hellenic Studies.
Kl—Klio.
MB—Musée Belge.
NGG—Nachrichten von der Gesellschaft der Wissenschaften zu Göttingen.
PBerol—Berlin Papyri.
Ph—Philologus.
POxy—Oxyrhynchus Papyri, ed. B. F. Grenfell and A. S. Hunt.
RE—Pauly-Wissowa-Kroll-Mittelhaus, Real-Encyclopädie der classischen Altertumswissenschaft.
RFIC—Rivista di Filologia e d'Istruzione Classica.
RPh—Revue de Philologie, d'histoire et de littérature anciennes.
RhM—Rheinisches Museum für Philologie.
SIG³—Dittenberger, Sylloge Inscriptionum Graecarum, Editio tertia.
TAPhA—Transactions of the American Philological Association.
WS—Wiener Studien.

*F. and T. are used to denote fragments and *Testimonia* in Jacoby's *FGrH*, Fg. those in Müller's *FHG*.

CHAPTER I

THE *Atthis* OF HELLANICUS

It is a hard task to reconstruct the development of early Greek historiography, since all the historical works that were written in the sixth and fifth centuries have been lost except for the histories of Herodotus and Thucydides. It is equally difficult to estimate how much these two authors owed to their contemporaries and immediate predecessors. Both of them, following no doubt the custom of their day, and unwilling to give a free advertisement to rivals in their own field, are disappointingly silent about contemporary writers. Hecataeus belonged to a previous generation, and Herodotus could therefore safely mention his name as author of two almost classical works. But his other allusions to historians and geographers are carefully vague; he speaks only of "the Ionians" or "the Greeks" or "certain Greeks anxious to gain a reputation for cleverness." [1] Thucydides mentions Hellanicus once, in order to remark that his *Atthis* was not detailed enough nor sufficiently exact in its chronology in dealing with the period between the Persian and Peloponnesian Wars. This is the only occasion, however, on which he mentions any historian by name; though he finds fault with several statements of Herodotus, he never actually names him. [2]

For further information about the progress of historical writing in the fifth century we are thrown back on much later authorities. From them we learn, what we might have suspected from the allusions in Herodotus, that most of these early historians came from the Aegean islands or from Asia Minor: Dionysius, Cadmus, and Hecataeus from Miletus; Hellanicus from Lesbos; Stesimbrotus from Thasos; Charon from Lampsacus; Damastes from Sigeum. At the same time the career of Herodotus is sufficient evidence that Athens could attract the Ionian historian and draw him away from his native city, just as surely as it attracted the sophists. But if

[1] E.g. 2.16, and 20; 3.32; 6.134; 7.151.

[2] Hellanicus is mentioned in 1.97 (see note 7 below); two passages of Herodotus are referred to in 1.20.3, as examples of mistaken ideas held by Greeks outside of Athens (οἱ ἄλλοι Ἕλληνες).

1

an Ionian historian was to make a name for himself in Athens, he could not confine himself to Ionian themes; he would be expected to apply his Ionian historical technique to questions which particularly interested Athenian readers.[3] These considerations explain in some degree why Hellanicus, a native of the Aeolian island of Lesbos, who was practised in the Ionian method of historical inquiry (ἱστορίη), was the first writer to attempt a specifically Athenian history—an *Atthis*.

Hellanicus was a prolific writer who wrote extensively about mythology, about the history and the country of the Persians and Egyptians, and about the early migrations of Greek tribes into Asia Minor and the islands. The exact number of his works is a problem which does not concern us here and seems insoluble in any case.[4] But a fortunate remark by a scholiast on the *Frogs* of Aristophanes gives us the information that his *Atthis* was not written or at least not completed until the closing years of the Peloponnesian War. This scholiast quotes Hellanicus as saying that the slaves who fought on the Athenian side in the battle of Arginusae were given their freedom and grouped with the Plataeans as Athenian citizens.[5] Some earlier scholars (for reasons which will be discussed later on) tried to evade the evidence of this scholion, but more recent criticism accepts it as a certain indication of the date of composition of the *Atthis*.[6] Thus the notion of an *Atthis* or local Attic history, as a literary form, was quite a new thing when Thucydides wrote, as indeed we might have guessed from his own words: "All my predecessors have neglected this period (the Pentecontaetia) and have dealt either with Hellenic affairs prior to the Persian Wars or with the Persian Wars themselves; the only author who has dealt with it at all, Hellanicus in his *Attic History*, has written too briefly and with too little accuracy in the matter of dates." [7]

---

[3] Note how readily Herodotus reports insulting remarks about the Ionians and affects to despise them, while admitting the advantages of their climate; cf. 1.142–43 (see note in How and Wells); 4.142; 5.69.

[4] Jacoby, *RE s.v.* "Hellanikos" 111–12, wisely refuses to spend much time on this problem which so much exercised earlier critics like Preller, Gutschmid (*Kl. Schriften* 4.316–26), and Kullmer. See bibliographical note at end of chapter. Cf. my *Early Ionian Historians* (Oxford, 1939), 155–56.

[5] Sch. Ar. *Ra.* 694: τοὺς συνναυμαχήσαντας δούλους Ἑλλάνικός φησιν ἐλευθερωθῆναι καὶ ἐγγραφέντας ὡς Πλαταιεῖς συμπολιτεύσασθαι αὐτοῖς, διεξιὼν τὰ ἐπὶ Ἀντιγένους τοῦ <πρὸ> Καλλίου. The discussion which follows will make it clear that this reference must be to the *Atthis*. Cf. esp. pp. 24–25.

[6] Jacoby, *loc. cit.* 109.

[7] 1.97: ἔγραψα δὲ αὐτὰ καὶ τὴν ἐκβολὴν τοῦ λόγου ἐποιησάμην διὰ τόδε, ὅτι τοῖς πρὸ ἐμοῦ ἅπασιν ἐκλιπὲς τοῦτο ἦν τὸ χωρίον, καὶ ἢ τὰ πρὸ τῶν Μηδικῶν Ἑλληνικὰ ξυνετίθεσαν

Hellanicus was the first to publish an historical work treating exclusively of Athenian affairs. Earlier writers had chosen a broader field and written *Hellenica;* Herodotus had not confined himself to Hellenic affairs, but had combined *Hellenica* with *Persica*, *Aegyptiaca*, *Scythica*, *Libyca*, and the affairs of other barbarian nations. Hellanicus also had worked in this more extensive field, probably in his younger days. But he was not the first to make an attempt at local history. Indeed local history was known as a literary form under the technical name of Ὧροι at least as early as Charon of Lampsacus; [8] and behind him again is the misty figure of Dionysius of Miletus, about whom so little is known and so much has been conjectured. [9] The date of Charon, as of many of the logographers, cannot be established with certainty. Suidas, as usual, is untrustworthy; the only really satisfactory piece of evidence comes from Plutarch, who in the *Life of Themistocles* [10] mentions Charon as one of the authorities who assumed that Artaxerxes was king when Themistocles took refuge at the Persian court. Since Artaxerxes succeeded Xerxes in 464, this remark provides at least a useful *terminus post quem* for the work of Charon, but for a *terminus ante quem* we must be content with the statement in the *De malignitate Herodoti* that Charon was "older than Herodotus." [11] In general, it is probably true to say that Ὧροι began to be written in Ionia some time later than 450 B.C.; it is only fair to leave Dionysius of Miletus out of the picture. This date, moreover, is a most suitable one for a revival of interest in Ionian local history. By the middle of the century the Greek cities of Asia Minor had recovered their independence from Persia; and with the recovery of independence and the renewal of local patriotism, it was natural that people should once again become interested in the past history of their native cities.

The middle of the fifth century also marks other developments in the Greek world that had a profound influence on Greek historical writing. The cessation of hostilities with Persia restored

ἢ αὐτὰ τὰ Μηδικά· τούτων δὲ ὅσπερ καὶ ἥψατο ἐν τῇ Ἀττικῇ ξυγγραφῇ Ἑλλάνικος, βραχέως τε καὶ τοῖς χρόνοις οὐκ ἀκριβῶς ἐπεμνήσθη.

[8] Suid. *s.v.* Χάρων Λαμψακηνός; Ath. 11.475B; 12.520D. Jacoby, "Ueber die Entwicklung der griech. Historiographie," *Kl* 9 (1909) 110–19, regards the development of Ὧροι as post-Herodotean. Laqueur's article in *RE s.v.* "Lokalchronik" is concerned more with the origin of official records than with the birth of a literary form.

[9] Cf. e.g. Ed. Meyer, *Forschungen* 1.176.

[10] 27.

[11] 859B. Cf. Tertullian, *Anim.* 46, who speaks of Charon as *Herodoto prior*. For a fuller discussion of Charon see *Early Ionian Historians*, chap. 4.

contact with the East and renewed the interest of Ionian readers in the manners and customs of the various nations under Persian rule. Two generations previously Hecataeus, in his *Periegesis* or Γῆς Περίοδος, had written on these subjects and now we find Herodotus and Hellanicus following his lead, continuing the geographical and ethnographical discussion which he first popularized. But if one of the literary forms developed by Hecataeus regained its popularity in works like the *Persica*, *Aegyptiaca*, and *Barbarica Nomima* of Hellanicus, mythography, the other branch of prose writing which he developed, suffered a different fate. The treatment of mythological themes by the tragedians inevitably overshadowed the work of prose mythographers, and later writers strove to attract attention by novelty and rationalistic discussion. Traces of this tendency are clear enough in the mythographical works of Hellanicus and become more pronounced later with Herodorus of Heraclea. But the more straightforward treatment of mythological subjects in prose, literary or instructional in character rather than controversial, naturally declined as tragedy developed. There is no certain proof that Hellanicus wrote his *Phoronis*, *Deucalioneia*, and *Troica* early in his career; but it would be in keeping with literary and historical developments if mythography occupied him in his earlier years and he devoted himself to ethnology and local history in the middle and latter part of his life.

With these considerations in mind we are in a position better to evaluate the evidence about the life of Hellanicus. Suidas, however untrustworthy, cannot be ignored.[12] He tells us that, in company with Herodotus, Hellanicus was at the court of the Macedonian king Amyntas "in the time of Euripides and Sophocles." There is some confusion here, since Amyntas I is too early for these dramatists and Amyntas II too late; but the statement that he went to the Macedonian court need not be dismissed because it is unintelligently reported. Suidas continues: "His career follows upon that of Hecataeus (since he was born at the time of the Persian Wars or even earlier)[13] and extends down to the time of Perdiccas;

---

[12] *S.v.* Ἑλλάνικος : Μιτυληναῖος, ἱστορικός, υἱὸς Ἀνδρομένους, οἱ δὲ Ἀριστομένους, οἱ δὲ Σκάμωνος· οὗ ὁμώνυμον ἔσχεν υἱόν. διέτριψε δὲ Ἑλλάνικος σὺν Ἡροδότῳ παρὰ Ἀμύντᾳ τῷ Μακεδόνων βασιλεῖ, κατὰ τοὺς χρόνους Εὐριπίδου καὶ Σοφοκλέους· καὶ Ἑκαταίῳ τῷ Μιλησίῳ ἐπέβαλε, γεγονότι κατὰ τὰ Περσικὰ καὶ μικρῷ πρός. ἐξέτεινε δὲ καὶ μέχρι τῶν Περδίκκου χρόνων, καὶ ἐτελεύτησεν ἐν Περπερήνῃ τῇ κατ' ἀντικρὺ Λέσβου. συνεγράψατο δὲ πλεῖστα πεζῶς τε καὶ ποιητικῶς.

[13] I have adopted Jacoby's emendation: γεγονώς . . . ἢ μικρῷ πρόσθεν for the MS. reading: γεγονότι . . . καὶ μικρῷ πρός.

he died at Perperene, on the mainland facing Lesbos; he wrote a large number of works both in prose and verse." This indication of date, consistent with the statement of Pamphila that he was born in 496,[14] is hard to reconcile with the scholion already quoted, which indicates that the *Atthis* was not composed till after 406; more especially since a catalogue of long-lived men [15] allows him a life of only eighty-five years. Either the tradition of Suidas and Pamphila is incorrect, or else the scholion, as it stands, must be disregarded. Diels and Rutherford chose to emend the text of the scholion: Diels substituted Θεόπομπος ἐν Ἑλληνικοῖς for Ἑλλάνικος; [16] Rutherford thought that the scholiast cited Hellanicus for the enfranchisement of the slaves who fought at Salamis, and that it was Philochorus who described the repetition of this reward after Arginusae.[17] More recent criticism has rejected these emendations, and reasonably so.[18] We do not know on what evidence the biographical tradition about Hellanicus rests; hence we are not justified in accepting its testimony, when that involves rejecting the clear and intelligible reading of the text of the Aristophanes scholiast, who is quoting an actual passage from the *Atthis*.

Pamphila's dates for the three historians are generally held to rest upon calculations in which the year of each author's *floruit* is taken to be his fortieth year: Thucydides is thought to "flourish" at the opening of the Peloponnesian War, so that the date of his birth is 471; Herodotus likewise is said to be born in 484, forty years before the colonization of Thurii, in which he took part; but why should the *floruit* of Hellanicus be 456? The most ingenious explanation is that of Rühl,[19] who thinks that Apollodorus in his *Chronica*, where Pamphila found her information, made Hellanicus a contemporary of Euripides; and that 456 was taken as the year of his *floruit* because Euripides produced his first tragedy in that year. Other ancient writers believed the two writers were both born on

---

[14] Gell. 15.23: Hellanicus initio belli Peloponnesiaci fuisse quinque et sexaginta annos natus videtur, Herodotus tres et quinquaginta, Thucydides quadraginta. Scriptum est hoc in libro undecimo Pamphilae.

[15] Luc. *Macr.* 22.

[16] *RhM* 31 (1876) 53.

[17] The scholion is printed as follows in his edition: τοὺς Σαλαμῖνι ναυμαχήσαντας δούλους Ἑλλάνικός φησιν ἐλευθερωθῆναι καὶ ἐγγραφέντας ὡς Πλαταιεῖς συμπολιτεύσασθαι αὐτοῖς· <καὶ Φιλόχορος > διεξιὼν τὰ ἐπὶ Ἀντιγένους τοῦ <πρὸ > Καλλίου <   >.

[18] For a strong protest against Diels' method see Wilamowitz, *H* 11 (1876) 291–94.

[19] *RhM* 61 (1906) 473–76.

the day of Salamis and that the name of Ἑλλάνικος, a contracted form of Ἑλλανόνικος, celebrated this "victory of the Greeks." [20]

It seems impossible to establish the date of Hellanicus' birth. But the character of his work evidently led critics to regard him as belonging to the generation before Thucydides; Dionysius of Halicarnassus in his letter to Pompey groups Charon and Hellanicus together as predecessors of Herodotus, though in his essay on Thucydides he speaks of Hellanicus and Herodotus as older contemporaries of Thucydides.[21] For the present, however, we are concerned principally with the date of the *Atthis;* whatever the exact date of his birth may be, we can accept the scholion as evidence that the *Atthis* was not published till the closing years of the fifth century.

About Hellanicus' family circumstances and his way of life we have no direct evidence. But we are entitled to draw some conclusions from the fact that he wrote an *Atthis.* Like Herodotus, he evidently abandoned his Ionian public and went to Athens in search of fresh success. We are much better informed about the literary tastes of Athens during the second half of the fifth century than about those of Ionia. The development of Athenian democracy had stimulated interest in oratory, and this was the field in which Athenian prose writers made most progress. This interest in the art of speaking encouraged some sophists to try the experiment of giving lectures on historical themes. Though Herodotus and Hippias of Elis are the only ones recorded to have been successful in this venture,[22] Jacoby maintains, plausibly enough, that Hellanicus had a successful career as a travelling lecturer.[23]

Ionian lecturers and historians, probably contemptuous of Athens as a state with a shorter artistic history than Ionia and jealous of its newly-won literary fame, would naturally pick out themes likely to appeal to popular taste. In the middle of the century the obvious theme was the struggle of the Greeks against Persia (Τὰ Μηδικά). But after 450 the interest in this subject waned. Tradition places the successful lectures of Herodotus prior to his departure for Thurii. With him we may suppose that treatment of Τὰ Μηδικά reached its highest pitch. At least, the tale of his enormous fee for a single lecture is an indication that his treat-

---

[20] T.6 (*FGrH* 1, no. 4)—*Vit. Eurip.* (ed. Schwartz) p. 2, 5. Whatever its historical value may be, this statement seems to prove that the penultimate syllable of his name is long.

[21] *Pomp.* 3, 7; *Th.* 9.

[22] Plu. *Malign. Herod.* 862A–B; Pl. *Hp. Ma.* 285D.

[23] *RE s.v.* "Hellanikos" 106.

ment was regarded as classical and final.[24]   What other subject
subsequently occupied the attention of historians?   Thucydides
gives us the answer: "Hellenic affairs prior to the Persian Wars." [25]
This, however, is an enormous field, and the historian might succeed
or fail according to his choice of material.   An Athenian audience
could scarcely be roused to enthusiasm over the affairs of Ionian
cities.   It was interested in its own history, in the affairs of its own
ancestors.

As Athenian historical interest quickened, intelligent Athenians
became aware how little knowledge they could muster of their own
city's past.   Certain episodes, of course, were generally familiar—
the tales which Herodotus had used, anecdotes of Solon and
Peisistratus, the story of Harmodius and Aristogeiton.   But history
before Solon was a blank.   The legendary figure of Theseus was
popular enough, but, like King Arthur in England, he was not
brought into relation with history.   In contrast with the Ionian
cities, many of whom could boast of rich and eventful histories,
Athens must have felt humiliated.   A golden opportunity, therefore, presented itself to the courageous writer who could reconstruct
the Athenian past, bring its legends into relation with reality, and
prove to the world that there was such a thing as Attic history.
It was a difficult task, needing the experience of a man well versed
both in legendary and in historical lore.   The ideal person would
be a man who had done some work in elucidating the early history
of other states and had studied mythology from a rationalist point
of view.   Such a man was Hellanicus, a writer with an established
reputation.   It would really be not at all surprising if he was
actually invited to Athens by Pericles and engaged to write the
first history of the Athenians.

This conclusion may seem fanciful, more especially as we do not
possess the text of his *Atthis*.   But the fragments, scanty as they
are, are exceedingly informative, and the arguments of the last
paragraph depend upon what is revealed in them and by the text
of Thucydides.   There is one point, however, which should be
made clear before the fragments are approached: Hellanicus never
became an important literary figure in circles where style was
regarded.   The few specimens of his narrative style which have

[24] Plu. *Malign. Herod.* 862B.   Gilbert Murray, *Ancient Greek Lit.*, 135, prefers to
rationalize this story, thinking the reward is for some serious public service, i.e. collecting material about regions politically important to Athens.
[25] 1.97, cf. note 7 above.

been preserved are not distinguished in any way, except for their pedantic insistence on obvious detail and their repetition of nouns instead of using pronouns—a style which suggests a lawyer's painstaking care to prevent any possible misunderstanding.[26]   It is not surprising, therefore, that Hermogenes ranks him among the writers who, so far as he can tell, left no influence on Greek style, and Cicero lists him as one of the early Greek authors to whom history meant no more than *annalium confectio*.[27]   That he did not interest his literary contemporaries at Athens is clear from their failure to mention him; except for the solitary reference in Thucydides, there is no explicit mention of him and no certain allusion to him in any classical Attic author.   Their taste was for oratory, rather than for dry historical records.   His day was to come later, when a reaction set in against the over-rhetorical treatment of history and the cry was for facts rather than wordy argument.   To the pedantic antiquarians of Alexandria in their search for curious *aetia* his works were a mine of information.[28]

The fragments of the *Atthis* show clearly that Hellanicus applied to early Attic history and national legend the same methods of research which he had followed in other branches of history and mythology.   A few characteristic fragments from his other works will suffice to show what this method was.   Scholiasts frequently refer to him as authority for the parentage of a mythical character; he sometimes confirms but often differs from the version current among the poets.   For example, he said that Dardanus married Bateia, daughter of Teucer (F.24), a lady whose name is not found in classical authors; the mother of Priam, according to his account, was not Zeuxippe, as in Alcman, but Strymo (F.139); and he gave a complete list of the divine lovers of the Pleiades and their children (F.19a).   It is probable that many of these details are innovations of his own, since considerable ingenuity and originality were necessary if he was to offer a consistent and comprehensive account of heroic genealogy.   There were many inconsistencies to be resolved,

---

[26] The best examples are F.28 (a rationalist explanation of Achilles' fight with the Scamander) and F.79a (the migration of the Sicels under King Sicelus to Sicily, formerly called Sicania, which took its new name "from this Sicelus who also became king in it").

[27] Hermog. *Id.* 2.12 (p. 412, ed. Rabe)—Hellanic. T.15; Cic. *De Orat.* 2.53—T.14.

[28] The scholia on Apollonius of Rhodes and Lycophron are enlightening in this respect.   The interest of Callimachus in local Attic tradition is illustrated by his engaging Ister to write a book on Attic history.   His relationship to Hellanicus and the Ionian logographers is a subject worthy of special study.

since the story tellers who built their tales round an individual representative of a legendary family tried to involve him in as many famous events as possible and often contradicted tales already current about some other hero.

To introduce some order into this confusion had been one of the aims of Hecataeus, and he had expressed his dissatisfaction with the existing condition of mythology in his famous introductory remark that "the tales told by the Greeks are many and ridiculous."[29] Hecataeus wished to give a reasonable version according to his own interpretation (ὥs μοι δοκεῖ), hoping that it would come to be regarded as the standard account. In this respect he was a successor to Hesiod, who tried to establish a standard theogony. His contemptuous remark about the conflicting accounts was doubtless a partial reminiscence of the famous lines of Hesiod:[30]

"These were the words they spoke to me first, those goddesses holy,
Muses Olympian, daughters of Zeus who carries the aegis:
'Shepherds so rude and worthless, shame on ye, bellies and nought else.
We can tell many a tale untrue that resembles a true tale,
But, when we so desire, we can tell truth in our verses.'"

Homer and Hesiod, as Herodotus tells us, established the family relations of the gods,[31] and no argument about them could be entertained by later writers. Hecataeus attempted, without success, to gain similar recognition for his account of heroic family relations. Hellanicus also followed in the Hesiodic tradition, but he had to reckon with contemporary rivals engaged in mythography and was doubtless more modest in his pretensions.

The traditional way of reckoning the passage of time in mythological narrative was by counting generations. It has been argued that Hecataeus reckoned the generation as equivalent to forty years,[32] but Herodotus reckons three generations to the century[33] without hinting that this is unusual, and it seems fairly certain that this was the orthodox reckoning of Ionian ἱστορίη in the latter part

[29] F.1a (*FGrH* 1, no. 1): Ἑκαταῖος Μιλήσιος ὧδε μυθεῖται· "Τάδε γράφω, ὥs μοι δοκεῖ ἀληθέα εἶναι· οἱ γὰρ Ἑλλήνων λόγοι πολλοί τε καὶ γελοῖοι, ὡς ἐμοὶ φαίνονται, εἰσίν."

[30] *Th.* 24–28.

[31] 2.33: οὗτοι δέ εἰσι οἱ ποιήσαντες θεογονίην Ἕλλησι καὶ τοῖσι θεοῖσι τὰς ἐπωνυμίας δόντες καὶ τιμάς τε καὶ τέχνας διελόντες καὶ εἴδεα αὐτῶν σημήναντες.

[32] Cf. Ed. Meyer, *Forschungen* 1.153–88; A. R. Burn, "Dates in Early Greek History," *JHS* 55 (1935) 130–46; and D. W. Prakken, "Herodotus and the Spartan King Lists," *TAPhA* 71 (1940) 460–72.

[33] 2.142. Heracleitus reckoned thirty years (H. Diels, *Vorsokr.*[5] 22 A 19). Cf. H. Fränkel, "Heraclitus on the Notion of a Generation," *AJPh* 59 (1938) 89–91.

of the fifth century.   The only way of establishing a chronological basis for mythology was to construct parallel family trees for the various great families beginning with the divine ancestor of each one.

It was comparatively rare, however, for tradition to preserve a continuous list for as many as ten generations.   If there were not names enough available to cover so long a period of time, some explanation of this fault in the tradition had to be devised.   A convenient solution was to maintain that the same name had been held by more than one member of the family, and that the less distinguished bearer of the name had been forgotten and his deeds (if any) falsely attributed to his more illustrious namesake.   Such a procedure was particularly convenient in explaining contradictions in the tradition, when the current tales about some hero were not entirely consistent.   The fragments of Hellanicus show several examples of characters duplicated in this manner.   For example, in his *Persica* he maintained that there were two kings of the name of Sardanapalus,[34] one an active conqueror, the other luxurious and idle; this was his way of explaining the somewhat inconsistent stories of the wealth of Sardanapalus.   Again, in the *Phoronis* he was anxious to show that the Pelasgians originated in Argos and bore that name before they went to Thessaly; accordingly he distinguished Pelasgus I, the Argive founder of the line, son or grandson of Phoroneus, from Pelasgus II, who led his subjects to Thessaly.[35]

It also seems extremely likely that Hellanicus duplicated Oenomaus, in order to complete the genealogical tree of the descendants of Atlas.   In this case we may have recourse to the *Bibliotheca* of Apollodorus, that mythological handbook which sets out to clarify the genealogy of heroic families and most certainly contains a great deal of material from early mythographic writers.   Many of its details agree with fragments of Hellanicus and it is not necessary to repeat here the arguments which prove that its author used Hellanicus as one of his principal sources.[36]   In the *Bibliotheca* Helen, as daughter of Tyndareus, is seven generations removed from Atlas.[37]   So also are Priam and Anchises; and Priam's grandfather Ilus, the founder of Ilium, is distinguished from another Ilus who dies childless, son of Dardanus and brother of Erichthonius;

[34] F.63.

[35] F.4 and Jacoby's note: cf. D.H. *Ant. Rom.* 1.17 and 28.

[46] The connection between Hellanicus and the *Bibliotheca* was realized long ago by L. Preller, *Ausgewählte Aufsätze* 29–30.

[37] 3.10.3.   See also 3.10.4, and Sir J. G. Frazer's note (Loeb ed., vol. 2, pp. 20–21).

this Ilus I is great-uncle to Ilus II.[38]   Menelaus' grandfather Pelops married Hippodameia, daughter of Oenomaus; but Hellanicus, in his *Atlantis* or *Atlantica*, tells us that the daughter of Atlas, Sterope, had a son called Oenomaus.[39]   If this Oenomaus is the father of Hippodameia, Menelaus is only five generations removed from Atlas.   This creates a difficulty, since Priam, Anchises, and Helen all belong to the seventh generation after Atlas.   The obvious solution is to suppose that there are two persons called Oenomaus; and if one is grandfather of the other, the genealogies of Helen and Menelaus work out perfectly in the following manner:

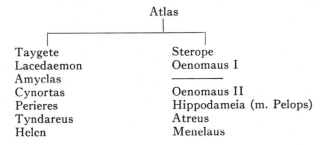

Apollodorus' genealogy of Priam and Anchises corresponds exactly with this:

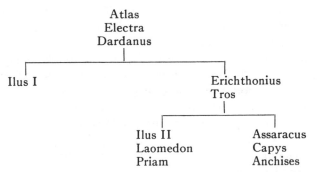

The *Bibliotheca* shows this same method in its list of Athenian kings;[40] the list is filled by using the same name more than once, and here too the hand of Hellanicus can be clearly seen.   Apollo-

[38] 3.12.1–3.

[39] F.19a.

[40] 3.14.5–15.5.   See also C. Frick, *Hellanikos von Lesbos und die Athenische Königsliste* (Progr. Höxter, 1880) and V. Costanzi, "L'opera di Ellanico di Mitilene nella redazione dei re Ateniesi," *Riv. di Storia Antica*, 8 (1904) 203–17, 243–53.

dorus, however, has not reproduced all the kings from Hellanicus' list, because he does not go back beyond Cecrops. His list of early kings is as follows: Cecrops, Cranaus, Amphictyon, Erichthonius, Pandion, Erechtheus, Cecrops II, Pandion II, Aegeus, Theseus. The duplication of the names of Cecrops and Pandion suggests the hand of Hellanicus. But to understand why in this case Hellanicus should have inserted new kings into the list we must examine what the chroniclers tell us of his chronological scheme.

Julius Africanus tells us that Hellanicus and Philochorus reckoned 1020 years from the time when Ogygus reigned in Athens (the time of the deluge) to the first Olympiad; [41] in other words, to use our modern system of dating for the sake of convenience, they gave 1796 B.C. as the date of Ogygus (presumably of his accession). The period from Ogygus to Cecrops, according to Philochorus,[42] was 189 years, and, in the absence of contradictory evidence, we may assume that here too, as in the date of Ogygus, he is following Hellanicus. Thus the reign of Cecrops begins in 1607, as compared with 1581 in the Parian Marble; [43] if 1796 marks the end, not the beginning of Ogygus' reign, a generation later, we can bring the date of Cecrops a little lower; but the earlier date, 1607, will do just as well. Now the list of kings in the *Bibliotheca* gives eleven generations from Cecrops to Demophon, the second successor of Theseus; and, according to Hellanicus, Troy was captured at the beginning of Demophon's reign.[44] Reckoning three generations to the century, we count 367 years from Cecrops' accession to Demophon's; and this gives us 1240 B.C. as the date for the fall of Troy— "about 800 years before my time," as Herodotus says,[45] evidently following the orthodox opinion of his day. Thus everything works out perfectly. · But if the list of kings were shortened, if Cecrops II and Pandion II were left out, the count of generations would be wrong. It will appear even more clearly later how necessary it was for Hellanicus to have exactly this number of kings from Cecrops to Demophon.

Another characteristic of his method is his use of etymologies and aetiological legends. Like Hecataeus before him, he delighted

---

[41] Hellanic. F.47a—Eus. *PE* 10.10 (488D). The form Ὠγυγος, rather than -ης, is preferable. Cf. J. Miller, *RE s.v.* "*Ogygos.*"
[42] Fg. 8 (*FHG* 1.385).
[43] Section 1 (*FGrH* 2.993).
[44] F.152a.
[45] 2.145.

to derive place names from mythological characters: thus Batieia, "a high place of Troy land," was named after Bateia, wife of Dardanus (F.24); Acele, a city in Lydia, after Acelus, a son of Heracles and Omphale's slave girl Malis (F.112). He also liked to explain the names of mythological characters: Pelias, according to his account, received his name from the livid mark on his face (ἐπεὶ ἐπελιώθη αὐτῷ ἡ ὄψις) where he had been kicked by a horse in his infancy (F.123);[46] and as for the iron-smiths of Lemnos, who nursed Hephaestus after his famous fall from heaven, they were called Sinties (so he said) because of "their acts of aggression against their neighbours" (παρὰ τὸ σίνεσθαι τοὺς πλησίον), since they were the first to manufacture offensive weapons of war (F.71c). Likewise in the *Atthis* he tells us that Munychia was named after an Athenian king Munychus (otherwise unknown), son of Pantacles (F.42); that the shrine of Artemis Colaenis was so called because it was founded by a certain Colaenus (F.163); and that the Areopagus got its name because Ares fixed (ἔπηξε; root, *pag.*) his spear in the ground there (F.38).

Evidently, then, in writing the *Atthis* Hellanicus did not abandon the devices which characterized his earlier mythological works. He applied his old methods to fresh material. With this preliminary conclusion established, the next step is to see what special peculiarities distinguish his *Atthis* from his other works, and how many of these peculiarities become established as part of a literary tradition.

The first and most striking feature of the *Atthis* is that it sets out to cover the whole of Attic history *ab urbe condita* up to the author's own time. This does not mean that it approximated in length to Livy's history of Rome or the universal histories of the fourth century. The fragments give some indication of its length, but unfortunately the evidence is incomplete. There are four fragments cited from the first book,[47] three of which refer to the period before Theseus. Of those from the second book, F.42 explains the origin of Munychia: how the inhabitants of Minyan Orchomenus, driven from their country by Thracians, came for refuge to Athens, and King Munychus allowed them to settle in the region which was called Munychia in his honour. One would expect this incident of heroic times to occur in Book I; since Book II is specifically

---

[46] Cf. Apollod. 1.9.8.
[47] F.38–41.

cited for the fragment, it must be assumed that the story was told in a digression. F.45 refers to the *Hierophantae* and F.46 to the *heroon* of Stephanephorus; such discussions of Attic institutions and cults might come at any point in the narrative. F.43 refers to Hippothoon, son of Alope and Poseidon, eponymous hero of the tribe Hippothontis, whose name would naturally appear in a description of the new tribal system established by Cleisthenes.

There remains F.44. Harpocration in his note on Pegae cites "Hellanicus in the fourth book of the *Atthis*." The earlier critics thought that Pegae would most naturally be mentioned in an account of the Pentecontaetia or the Peloponnesian War, and that the fourth book was the closing book. This reference to a fourth book stands alone, however, and Jacoby will not accept it as genuine, arguing that the *Atthis*, like the other major works of Hellanicus, except the *Priestesses*, consisted of two books only.[48]

If there are not more than four books, the treatment of the mythical period may seem disproportionately long. But tradition about early times was abundant, whereas for the later period, as far as the fifth century, it was almost wholly lacking except for certain episodes, like the tales of Cylon and Draco, Solon and Cleisthenes, and the rise and fall of the tyrants. In the fifth century itself, Hellanicus' account of the Pentecontaetia was not detailed enough to satisfy Thucydides. On the other hand, when he reached the Peloponnesian War and events which had taken place after his own arrival in Athens, he was in a position to offer much more detailed treatment. The scholion on the *Frogs* already quoted [49] refers to him as describing "the events in the archonship of Antigenes"; and another scholion on the same play remarks that "Hellanicus records the minting of a gold coinage in the archonship of Antigenes." [50] The implication is that he described the events of these latter years in true annalistic fashion; that the eponymous archon of each year was mentioned and the events during his term recorded. Such, indeed, as will appear, was the method of Philochorus for events in the fourth century; and the method of Thucydides is similar. The question is: at what point in Attic history did Hellanicus introduce this method? The criticism of Thucydides precludes the possibility that he used it for the Pentecontaetia.

[48] Cf. his remarks in *RE* s.v. "Hellanikos" 142.

[49] See note 5 above.

[50] F.172—Sch. Ar. *Ra.* 720, with Bentley's emendation Ἑλλάνικος for the meaningless ἀλλὰ νικᾷ.

If we could establish the fact that he gave an annalistic account for the whole period of the Peloponnesian War, the result would be most important for the criticism of Thucydides and the first two books of Xenophon's *Hellenica*. But a definite conclusion is impossible.

It seems, however, that the subject matter of the *Atthis* falls naturally into three divisions, and that each division was treated in a different way: the period of the kings, the historical period up to the middle of the fifth century, and the period of the Peloponnesian War. The first book is not cited for any event later than the time of Theseus and it is commonly believed that Book I dealt only with the regal period. The fragments themselves and the division of the material by later Atthidographers show how much attention was devoted to this mythical age. Hellanicus' method of dealing with it deserves closer attention.

Starting at the beginning, he would be obliged to establish, or at least to assert, the claim of the Athenians to be autochthonous. Harpocration quotes Hellanicus for the statement that the Arcadians, Aeginetans, and Thebans were autochthonous: [51] he does not tell us where this statement was made, but it seems most likely that it came in the *Atthis* and as a comment on the Athenian claim. Very few traditions survive about the kings before Cecrops, the immediate successors of Ogygus. Munychus, who gave his name to Munychia, is presumably one of these early kings,[52] and Colaenus is another. A scholion on the *Birds* gives the comment: "Hellanicus says that Colaenus, a descendant of Hermes, established a temple of Artemis *Colaenis* in obedience to an oracle." [53] Pausanias was interested in this shrine and mentions the tradition that Colaenus ruled in Athens before Cecrops, as well as the tradition that he led some settlers to Colonides in Messenia; [54] it is likely enough that both pieces of information came from Hellanicus.

The period from Cecrops to Demophon was marked by four famous trials before the Areopagus: the trial of Ares, opposed by Poseidon, for killing Halirrhothius, of Cephalus for killing Procris, of Daedalus for killing Talos, and finally the trial of Orestes. The *Eumenides* had aroused interest in the early history of the Areopagus, and it appears that Hellanicus expected to find this interest

[51] *S.v. αὐτόχθονες* (F.161).
[52] F.42. See p. 13 above.
[53] Sch. Ar. *Av.* 873 (F.163).
[54] 1.31.5; 4.34.8.

still alive, perhaps renewed by the oligarchic agitation after the Sicilian disaster. At the time of the democratic movement in the sixties, when the Areopagus was shorn of its political powers and survived only as a court for homicide trials, it was natural that historical precedent for its judicial activity should be sought. Aeschylus captured popular imagination by his treatment of the trial of Orestes; but it is from the fragments of Hellanicus that we first learn the details about earlier trials. Hellanicus is the first authority to record in full how Ares was tried for killing Halirrhothius. Allusions in the *Electra* and the *Iphigeneia in Tauris* of Euripides [55] show that he did not invent the tale, but it finds no mention in earlier literature.[56] How Cephalus killed his wife Procris was told in full by Pherecydes;[57] but there is no earlier authority known for the story and for the trial of Cephalus no authority before Hellanicus.[58] So also for the trial of Daedalus, who killed his nephew and pupil Talos, Hellanicus is the earliest known authority.[59]

The absence of these tales from earlier literature suggests that they were invented for political purposes in the middle of the fifth century. But it was still open to Hellanicus to improve and rationalize them and put them in their chronological setting. According to his account, as reported by a scholiast on the *Orestes* of Euripides,[60] the trials took place at intervals of three generations. At first sight there seems to be a difficulty in the account. Halirrhothius is killed by Ares for attempting to rape Alcippe, granddaughter of Cecrops, an event which should most probably be placed in the reign of Cecrops' successor, Cranaus; Apollodorus describes the trial in his account of the reign of Cecrops, the Parian

[55] *El.* 1258–63; *IT* 945–46.
[56] Cf. P. Friedländer, *RE* s.v. "Halirrhothios" 2268–69.
[57] F.34 (*FGrH* 1.71).
[58] Cf. Schwenn in *RE* s.v. "Kephalos" (1) 218–19. See also M. A. Schwartz, *Erechtheus et Theseus apud Euripidem et Atthidographos* (Leiden, 1917) 39–43.
[59] For later references see van der Kolf in *RE* s.v. "Talos" 2087.
[60] There are two separate scholia. The first (1648—F.169a) gives the actual words of Hellanicus, but the text is exceedingly corrupt; the second (1651—F.169b) sums it up as follows: πάγοισιν ἐν Ἀρείοισιν. ἐνταῦθα πρῶτον μὲν Ἄρης καὶ Ποσειδῶν ἠγωνίσαντο· δεύτερον δὲ μετὰ τρεῖς γενεὰς Κέφαλος ὁ Δηιονέως ἐπὶ γυναικὶ Προκριδι· καὶ μετὰ τρεῖς Δαίδαλος ἐπὶ τῷ ἀδελφιδῷ Τάλῳ· εἶτα μετὰ τρεῖς Ὀρέστης, ὡς Ἑλλάνικος. Max Wellmann, "Beitrag zur Geschichte der attischen Königsliste," *H* 45 (1910) 554–63, thinks that this scholion argues a shorter list of kings than that of Apollodorus, but he is surely wrong in taking μετὰ τρεῖς γενεάς as equivalent to τρίτη γενεᾷ ὕστερον, i.e. *two* generations later. See also G. De Sanctis, Ἀτθίς (ed. 2) 114–16.

Marble puts it in the reign of Cranaus.[61]   The trial of Orestes must take place in the reign of Demophon, about fifteen years after the fall of Troy, and this is ten, not nine, reigns later than the reign of Cranaus.   This difficulty, however, disappears when one remembers that each reign need not correspond exactly with the length of a generation and that Amphictyon, Cranaus' successor, reigned only twelve years.[62]   The trial of Cephalus, therefore, might well be put in the reign of Erechtheus, four reigns after Cranaus, and rather more than a hundred years after the first trial; not precisely three generations or 100 years, but at all events less than four generations.   The chronological scheme of Hellanicus, then, would be as follows (the dates of course are approximate, and no great importance should be attached to them, except to the date of the fall of Troy):

1. Cecrops—accession 1607 B.C.
2. Cranaus—trial of Ares and Poseidon—c.1550
3. Amphictyon (short reign)
4. Erichthonius
5. Pandion
6. Erechtheus trial of Cephalus—c.1440
7. Cecrops II
8. Pandion II
9. Aegeus—trial of Daedalus—c.1330
10. Theseus
11. Menestheus—fall of Troy—1240
12. Demophon—trial of Orestes—c.1230–25

The *Bibliotheca* of Apollodorus mentions various incidents supposed to have taken place in the period before Theseus, some of which were doubtless recorded by Hellanicus.   All earlier characters, however, are insignificant as compared with Theseus himself.   Plutarch's *Theseus* shows how much attention was devoted to this hero not only by Hellanicus but also by the later Atthidographers.   It is noteworthy indeed that Plutarch refers to the Atthidographers for his evidence, rather than to the dramatic and lyric poets.   The place subsequently occupied by Theseus in the

[61] 3.14.2; *Mar. Par.* section 3.
[62] 3.14.6.   It is interesting to note that Herodotus gives a reign of only twelve years to Sadyattes, the third Mermnad king of Lydia (1.15); this short reign offsets the long reigns of Ardys and Alyattes and the total reigns of the five kings amount to 170 years—a proper period for five generations.   He does not try to achieve such a perfect result for the earlier Heraclid kings, whose 22 reigns ('with son succeeding father') cover only 505 years.

tradition of the *Atthis* will be discussed in a later chapter. But the fragments reveal that Hellanicus was instrumental in establishing a tradition about Theseus, and especially in attributing to him certain characteristics which made him into an Athenian Heracles. His work on *Ktiseis* or *Foundings of Cities* had doubtless taught him something about the popular attitude towards national heroes, and in the *Phoronis* he had dealt with the legend of Heracles. Theseus was, in a sense, the founder of Athens since he was responsible for the synoecism, and a most suitable character for a national hero. Was he recognized as such before Hellanicus elevated him to this rank?

Reverence for Theseus as an οἰκιστής doubtless goes back to early times.[63] But it is true that the actual ἆθλοι of Theseus are not attested before the fifth century. The scanty references to Theseus in early literature mention only his slaying of the Minotaur and his abduction of Ariadne and Helen.[64] The evidence of works of art is similar. Apart from the Cretan adventures, some black-figure vases show him fighting against the centaurs; only *one* shows his conflict with the Amazons, although their fight with Heracles was frequently portrayed. His struggle with the Marathonian bull is difficult to identify, because of this animal's resemblance to the Minotaur. The other labours appear for the first time in works of the fifth century—for example, in the metopes of the so-called Theseum at Athens and on red-figure vases. Probably, therefore, it is fair to say that, though these tales were well enough known in the time of Hellanicus, much remained to be done by way of rearrangement and addition of detail in order to establish a definite Attic tradition of his life. In this task, Hellanicus' experience as a mythographer was bound to prove itself useful.

[63] De Sanctis, however ('Aτθίς 22–24), thinks the tradition of the synoecism of Theseus is of quite late origin; that Thucydides is theorizing on his own, rather than recording what is generally accepted.

[64] F. H. Wolgensinger, *Theseus* (Diss. Zürich, 1935) 7–9, cites Homer, *Il.* 1.265; *Od.* 11.322, 631; Alcman, Stesichorus (Paus. 1.41, 4, 2.22, 6), and Sappho (Serv. *Aen.* 6.21). He also gives the evidence from works of art. Similar conclusions about the date of origin of the labours of Theseus were reached over sixty years ago by W. Gurlitt, *Das Alter der Bildwerke und die Bauzeit des sogenannten Theseion in Athen* (Vienna, 1875). Cf. also Preller-Robert, *Griech. Mythologie* (Berlin, 1921), 2.2.676–756, and H. Herter, "Theseus der Ionier," *RhM* 85 (1936) 177–91, 193–239, and "Theseus der Athener," *RhM* 88 (1939) 224–86, 289–326. Herter's main thesis, that the glorification of Theseus took place in Peisistratid times is not adequately supported by the evidence. A new black-figure vase from the Athenian agora, apparently depicting one of the combats of Theseus, is reported by T. L. Shear, *Hesperia* 8 (1939) 229–30.

The fragments show that among the exploits of Theseus described by him were the slaying of the Minotaur, the founding of the Isthmian games, the expedition against the Amazons, the victory in Attica over the Amazons, and the abduction of Helen and Persephone, the latter apparently from the lower world.[65] These are exploits parallel and comparable to the labours of Heracles, which he treated at some length in the *Phoronis*.   Plutarch gives enough details to show the method of his treatment, which was evidently highly circumstantial.   The rape of Helen (especially since it involved a descent on the Peloponnese) could not be omitted, but there were chronological difficulties: Theseus was more than a generation older than Helen; [66] hence Hellanicus insists that Theseus was over fifty years old and Helen a child of seven at the time of her abduction.[67]   The legend of the Minotaur was more completely rationalized in the later tradition; [68] Hellanicus' work contains no trace of such rationalization, but instead an exact account of the circumstances which led to Theseus' departure for Crete: "Hellanicus says," writes Plutarch, "that the city did not send its young men and maidens by lot, but that Minos himself used to come and pick them out, and that he now picked out Theseus first of all, following the terms agreed upon; and he says the agreement was that the Athenians should furnish the ship, and that the youths should embark and sail with him carrying no weapons of war, and that the penalty was to be exacted no longer if the Minotaur were killed." [69]   If this account deprived Theseus of the credit for volunteering to go of his own accord, it at least magnified the wickedness of Minos and the despair of the Athenians who, if they went unarmed, could never hope to kill the Minotaur.

Up to the time of the Trojan War the chronological scheme rested upon counting generations.   This system could not be continued further, because tradition did not offer enough material upon which to build.   It might have been possible to continue the system until the death of Codrus, the last Athenian king; but if there was no traditional date for his reign, and no tradition about

---

[65] F.164–68—Plu. *Thes*. 17; 25; 26; 27; 31; F.134—Sch. Hom. *Il*. 3.144.

[66] The Trojan War breaks out in the middle of the reign of his successor.

[67] F.168a—Plu. *Thes*. 31: ἤδη δὲ πεντήκοντ' ἔτη γεγονώς, ὥς φησιν Ἑλλάνικος, ἔπραξε τὰ περὶ τὴν Ἑλένην, οὐ καθ' ὥραν.   F.168b—Sch. Lyc. 513: φησὶ δὲ ὁ Ἑλλάνικος ἑπταετῆ οὖσαν Ἑλένην ἁρπαγῆναι ὑπὸ Θησέως.

[68] Cf. Philoch. Fg.38–40 (*FHG* 1.390f.).

[69] F.164—Plu. *Thes*. 17.

his immediate predecessors, Hellanicus could scarcely venture to reconstruct with so little foundation. The same difficulty holds good for the early archontate; there is no way of telling exactly what records, whether genuine or falsified, were preserved in the *Metroon*, but a mere list of archons' names would be of little help to him.[70]

The lack of fragments about the early historical period and indeed the generally scanty information given in Athenian literature about the period from the ninth to the sixth centuries suggests a further conclusion about the *Atthis* of Hellanicus. Admittedly it is rash to conjecture the character of a work from the absence of references, and conclusive proof is impossible. But the negative evidence about the *Atthis* is borne out by similar evidence about the other Atthidographers, and by the text of Herodotus, Thucydides, and Aristotle's *Constitution of Athens*. Not only is it unlikely that Hellanicus had any chronological scheme for this period; it is unlikely that he even narrated many historical events from it, for the good reason that neither records nor tradition offered much material. In the lack of tradition about historical events, he would be obliged to fill the space somehow with material of local interest. Fragments of his own *Atthis* and of the later *Atthides* suggest very strongly that this material related to topography and to religious cults; and his experience in writing the *Ktiseis* would render him inclined to seek this way out of the difficulty.

There is no necessity here to emphasize how greatly the later Atthidographers were interested in religious material.[71] Philochorus was himself an Exegetes and is supposed to have written works on specifically religious subjects: *On Festivals*, *On Sacrifices*, *On the Soothsayer's Art*. Cleidemus and Phanodemus show a similar interest. It seems not to be generally recognized, however, that the tendency of the *Atthides* to deal with religious matters is the result of a tradition started by Hellanicus. Of the five fragments quoted from Hellanicus' second book two are concerned with religious institutions. Harpocration (*s.v.* Ἱεροφάντης) remarks: "Hellanicus has discussed the clan of the Hierophants in Book II of his

---

[70] De Sanctis ('Ατθίς, 99–116) has made an attempt, which is not entirely convincing, to show that Philochorus gave a complete list of Athenian rulers as far down as the founding of the annual archontate. He does not try to reconstruct Hellanicus' list of kings after Demophon.

[71] Cf. A. Tresp, *Die Fragmente der griech. Kultschriftsteller*, in *Religionsgesch. Versuche u. Vorarbeiten* 15.1 (Giessen, 1914).

Atthis";[72] and again (*s.v.* Στεφανηφόρος) he distinguishes Stephanephorus the son of Heracles (also mentioned by Hellanicus) from an
Attic hero of that name with a *heroon* "to which Hellanicus refers
in the second book of the *Atthis*."[73] Another fragment from the
second book speaks of the naming of the tribe Hippothontis,[74] a
matter of religious as well as of political importance. The other
two fragments concerning Pegae and Munychia have already been
discussed.[75]

The fragments of the *Priestesses* offer further evidence. The
Priestesses is supposed to have been a chronological work, which
dated various events in the Greek world by the name and year of
the priestess of Hera holding office at Argos. The elaborate chronological statement of Thucydides at the beginning of his second book
takes account of this method of dating: "When Chrysis had been
priestess in Argos for forty-eight years, when Aenesias was ephor
at Sparta and Pythodorus still had four months to run as archon
in Athens, etc." Although it is unnecessary to suppose that this
particular date is taken from Hellanicus' work, it is generally
admitted that the *Priestesses* dealt with events in the Peloponnesian
War. Stephanus in his note on Chaonia quotes some actual words
of Hellanicus in the third book of the *Priestesses:* "Ambraciots and
the Chaonians and Epirotes who accompanied them," and this
phrase is most naturally understood in reference to the campaign
in Ambracia of 429, described by Thucydides in 2.80–82.[76] Since
fragments of the first book of this work are full of references to
mythical characters like "Macedon, son of Aeolus," and "Phaeax,
son of Poseidon,"[77] it must have covered just as large a period as
the *Atthis;* but if the year 429 had already been reached in the
third book, some portions of the period were evidently treated in
a very perfunctory manner.

It appears, moreover, that the resemblance of the *Priestesses*
to the *Atthis* went further than this. The fragments show that it
contained a number of *aetia* and topographical notes, and of the
ten fragments cited from it only one, besides that already quoted,

---

[72] F.45.
[73] F.46.
[74] F.43—Harp. *s.v.* 'Αλόπη : . . . Κερκυόνος θυγάτηρ, ἐξ ἧς καὶ Ποσειδῶνος 'Ιπποθόων
ὁ τῆς 'Ιπποθοωντίδος φυλῆς ἐπώνυμος, ὡς 'Ελλάνικος ἐν β' 'Ατθίδος.
[75] F.44, 42. Cf. pp. 13–14 above.
[76] F.83: 'Αμβρακιῶται καὶ οἱ μετ' αὐτῶν Χάονες καὶ 'Ηπειρῶται.
[77] F.74, 77.

deals with an incident of historical times: the founding of Naxos in Sicily by Chalcidians and Naxians under the leadership of Theocles.[78] No doubt this event was dated in accordance with the traditions and records preserved in this Sicilian city. But the number of fragments not strictly chronological in character suggests that Hellanicus found the same trouble as in the *Atthis;* that there was insufficient material relating to chronology and that the gap had to be filled by references to myths and genealogy and the foundation legends of cities—such as the legend that Chaeroneia was founded by Chaeron, son of Apollo and Thero, and Corcyra was named after Cercyra, mother of Phaeax by Poseidon.[79] A date, however, had been fixed by some means or other, probably in Sicilian tradition, for the Sicel emigration to Sicily, in the twenty-sixth year of the priestesship of Alcyone.[80]

Furthermore, even if this argument about the character of the *Atthis* is in the main an *argumentum ex silentio,* based on the lack of fragments referring to historical incidents, the silence is not confined to Hellanicus alone. Both Herodotus and Aristotle's *Constitution of Athens* are lacking in information about the early Athenian historical period, except as regards a few incidents. The lost chapters at the beginning of Aristotle's work appear to have dealt with legendary and semi-legendary times; his treatment of the seventh century, which is substantially intact, is extremely meagre. Plutarch and Pausanias, despite their knowledge of the Atthidographers, can add very little of genuine historical value. Even on the historical questions which were popularly discussed, because they marked epochs in Athenian history, the variation among the accounts shows the lack of an authoritative version. Thucydides shows this lack by his protest against what he considers the current version of the tale of Harmodius and Aristogeiton; he

---

[78] F.82—St. Byz. *s.v.* Χαλκίs : . . . Ἑλλάνικος Ἱερειῶν Ἥρας β'· "Θεοκλῆς ἐκ Χαλκίδος μετὰ Χαλκιδέων καὶ Ναξίων ἐν Σικελίῃ πόλιν ἔκτισεν." F.81 is disputed. St. Byz. *s.v.* Χαιρώνεια tells how the town was named after Chaeron; τοῦτον δὲ μυθολογοῦσιν Ἀπόλλωνος καὶ Θηροῦς, ὡς Ἑλλάνικος ἐν β' Ἱερειῶν Ἥρας. Then follows an apparent quotation: Ἀθηναῖοι καὶ <οἱ> μετ' αὐτῶν ἐπὶ τοὺς Ὀρχομενίζοντας τῶν Βοιωτῶν ἐπερχόμενοι καὶ Χαιρώνειαν πόλιν Ὀρχομενίων εἷλον, which seems to refer to the Athenian campaign in Boeotia in 447, but does not follow naturally on what goes before. Jacoby accepts the existence of a lacuna. Wilamowitz (*Aristoteles u. Athen* 1.281 note) thinks the quotation may be from Theopompus.

[79] F.81, 77.

[80] F.79b—D.H. *Ant. Rom.* 1.22.3: τὸ μὲν δὴ Σικελικὸν γένος οὕτως ἐξέλιπεν Ἰταλίαν, ὡς μὲν Ἑλλάνικος ὁ Λέσβιός φησι, τρίτῃ γενεᾷ πρότερον τῶν Τρωικῶν, Ἀλκυόνης ἱερωμένης ἐν Ἄργει κατὰ τὸ ἔκτον καὶ εἰκοστὸν ἔτος.

finds it necessary to point out that Hipparchus, whom they killed, was not tyrant at the time, despite the evidence of the famous *scolion*:

καὶ τὸν τύραννον κτανέτην
ἰσονόμους τ᾽ Ἀθήνας ἐποιησάτην.[81]

Furthermore, epigraphic evidence is and apparently always was lacking. Quotations from and appeals to the so-called *axones* and κύρβεις of Solon, whatever their historical and legal value may be, prove nothing about the survival of the original wooden tablets which Solon is supposed to have set up in the Stoa Basileios.

At the same time, the rise of the Solonian legend is not itself an indication of historical ignorance; it is not necessary for a national hero, "the father of his country," to be an altogether obscure figure before legends can accumulate about him.[82] More significant is the fact that, among all the accusations made against the revolutionary oligarchs of the Four Hundred and the Thirty, the charge of misleading the people by falsifying history never occurs. The forging of a Draconian constitution appears to have been a comparatively safe political manoeuvre;[83] and the proposal of Cleitophon to "investigate the ancestral laws of Cleisthenes" seems to imply the lack of any authentic tradition about the details of his reforms.[84] From the divergent theories of the Atthidographers we can conclude that by the middle of the fourth century people had become aware that a great deal of "interpretation" was needed

[81] 6.54.

[82] This is one reason why Plutarch is not necessarily a less trustworthy authority for the work of Solon than Aristotle. For the prevailing ignorance in Athens about Solon's constitutional reforms, cf. I. M. Linforth, *Solon the Athenian* 74–76, 278–84; Ed. Meyer, *Geschichte des Altertums* (ed. 2) 3.600, note. ·

[83] This is the prevailing interpretation of the mysterious Draconian constitution, as described by Aristotle, *Resp. Ath.* 4; that it was a forgery, made by the oligarchs in order to supply historical precedent for some of the details in their suggested constitution εἰς τὸν μέλλοντα χρόνον (*Resp. Ath.* 30–31). Cf. Busolt, *Griech. Staatskunde* 52–58.

[84] Arist. *Resp. Ath.* 29.3: Κλειτοφῶν δὲ τὰ μὲν ἄλλα καθάπερ Πυθόδωρος εἶπεν, προσαναζητῆσαι δὲ τοὺς αἱρεθέντας ἔγραψεν τοὺς πατρίους νόμους οὓς Κλεισθένης ἔθηκεν ὅτε καθίστη τὴν δημοκρατίαν, ὅπως ἀκούσαντες καὶ τούτων βουλεύσωνται τὸ ἄριστον, ὡς οὐ δημοτικὴν ἀλλὰ παραπλησίαν οὖσαν τὴν Κλεισθένους πολιτείαν τῇ Σόλωνος. J. A. R. Munro, "The Ancestral Laws of Cleisthenes," *CQ* 33 (1939) 84–97, has offered a very remarkable explanation of this passage; he suggests that there is not merely *one* constitution of Cleisthenes but *three*, of which the first is undemocratic and reactionary; he thinks that the good reputation enjoyed by Cleisthenes in later times was largely due to the efforts of his kinsmen, the Alcmaeonidae, whose influence "shamelessly colours, conceals, or distorts the truth."

before the story of the origins of their constitution could be understood.

In brief, therefore, the argument may be summed up as follows.
No fragment of Hellanicus or of the other Atthidographers refers
to any important Athenian event in the early historical period
which is not recorded elsewhere.   The silence of other historical
authorities suggests very strongly that authoritative evidence about
this period was in the main lacking; and this makes us ready to
believe that the silence of the fragments is not merely fortuitous.
On the other hand, in a comprehensive work Hellanicus could not
jump from mythical times to his own day without making some shift
to treat of the intervening period.   The evidence of his fragments
and of the fragments of the later *Atthides* suggests that he evaded
the difficulty by substituting religious and topographical discussion
for historical narrative.   It should follow, therefore, that so far as
he established any tradition for the early historical period, this
was a religious rather than an historical tradition.   This question
will call for further discussion in later chapters.

It remains to examine the evidence relating to the third portion
of the *Atthis*, which dealt, as it appears, in more detailed fashion
with events of the Peloponnesian War.   The direct evidence is
confined to three fragments.   The scholion on the *Frogs*, referring
to the slaves who fought at Arginusae, has already been quoted.[85]
In another scholion on the same play it is mentioned that Hellanicus
recorded the minting of a gold coinage in the same year.[86]   In each
case it is made quite clear that Hellanicus dated the event in the
year of the archon Antigenes, and in the first scholion he is said to
report the enfranchisement of the slaves "when describing the
events in the archonship of Antigenes."   There is no explicit reference to the *Atthis* and the difference in character of these fragments
from others referred to the *Atthis* is marked.   But fragments of a
similar character from the *Atthides* of later writers (especially those
quoted from Philochorus by Didymus) seem to show that this kind
of annalistic record was customary in an *Atthis;* that it was one of
the traditional characteristics of any historical work written under
this title.   When so many other features well established in the
literary tradition can be traced back to Hellanicus, there is no good
reason for denying his influence in this particular detail.   Accord-

[85] F.171 (see note 5 above).
[86] F.172 (see note 50 above).

ingly, these two fragments from the scholia on Aristophanes may be taken as certain evidence that a portion of his *Atthis* was devoted to an annalistic record, with the events of each year grouped under the name of its eponymous archon.

Unfortunately there is no certain indication, either for Helianicus or the later Atthidographers, at what date their annalistic treatment began. Likewise, the uncertainty about the number of books in the *Atthis* makes it impossible to know what proportion of the work was occupied by this section. But at whatever point he started it, it marked a complete break from the earlier portion, and it is likely that the break was marked by the beginning of a new book.

It is worthy of note that Thucydides' history, like Hellanicus', falls into three divisions, an *Archaeologia*, a summary of historical events previous to the Peloponnesian War, and a detailed annalistic treatment of the later period. But in Thucydides the final portion is seven times as long as the first two parts put together. Hellanicus evidently observed a very different proportion and doubtless this was a feature of his work which irritated Thucydides.

Yet despite the difference in character of the final portion of the *Atthis* from the earlier part, Hellanicus did not exclude genealogical interests altogether from his account of the Peloponnesian War. This is shown by a third fragment referring to this period, which is reported in three separate places.[87] Andocides, the orator, according to Hellanicus, was descended from Hermes—a matter of considerable interest since he was suspected of complicity in the affair of the Hermae—but also from Telemachus and Nausicaa. The account in the *Lives of the Ten Orators* is muddled and misses the point of the genealogy: after mentioning his reputed descent from Hermes, it adds that for this reason he was sent with Glaucon to help the Corcyreans against the Corinthians in the conflict preceding the Peloponnesian War. Clearly Hellanicus gave his descent from Nausicaa as the reason; because Corcyra was supposed to be the ancient Phaeacia, and an Athenian of Phaeacian descent would be a most suitable emissary. Thucydides [88] mentions Andocides, son of Leogoras, as one of those in command of the ships sent to Corcyra, but his name does not appear in the Athenian inscription recording funds voted for this campaign.[89] Consequently Thucydides has

---

[87] F.170a—*Vit. X Orat.* 834B; F.170b—Plu. *Alc.* 21; F.170c—Suid. *s.v.* Ἀνδοκίδης.
[88] 1.51.
[89] M. N. Tod, *Gk. Historical Inscriptions* no. 55.

been suspected of a mistake.   In his defence it has been suggested that perhaps Andocides was only unofficially attached to the expedition,[90] though it is also possible that Dracontides (whose name is given in the inscription) resigned in his favour before the expedition started.   In any case, Andocides the orator would have been too young to be στρατηγός at the time, and his grandfather, who is likely to have borne the same name, would have been too old. Plutarch understands the statement to apply to the orator; but here of course he may be misunderstanding Hellanicus.   If, therefore, the appointment was an irregular one, the tale of his Phaeacian descent may be supposed to explain it.   Whether such a reason was really offered or not is immaterial.   The interest of the fragment is that it shows how Hellanicus retained to the end the interest in genealogy which he inherited from Hecataeus, and brought genealogical discussion into contemporary questions.   One of the tasks of the next chapter will be to show how far even Thucydides was influenced by this interest in genealogy, which was firmly established in the tradition of Ionian historiography.

BIBLIOGRAPHICAL NOTE

The fifth chapter of my *Early Ionian Historians* is devoted to Hellanicus and a portion of the chapter to his *Atthis*.   In the preceding discussion it has been necessary to go over some of the same ground and a certain amount of repetition has been inevitable.   The only alternative to such repetition would have been a poorly balanced argument with numerous cross-references.   Among earlier discussions of Hellanicus the following may be mentioned:

L. Preller, "De Hellanico Lesbio," in *Ausgewählte Aufsätze* (1864).
H. Kullmer, "Die Historiai des Hellanikos von Lesbos: ein Rekonstruktionsversuch,"
    Neue Jahrbücher für Philologie, Supp. 27 (1901) 445–691.
F. Jacoby, *RE s.v.* "Hellanikos."
W. Schmid, *Griechische Literaturgeschichte*, 1.2.680–92.
For a fuller bibliography see *Early Ionian Historians* (Oxford, 1939), 233–35.

The fragments are quoted according to Jacoby's edition in *Die Fragmente der griechischen Historiker* (referred to as *FGrH*) 1, no. 4.   The abbreviations T. and F. are used for *Testimonium* and *Fragmentum*.

[90] Cf. e.g. Jacoby's note on F.170; Tod, *loc. cit.*; G. F. Hicks, *Gk. Hist. Inscr.* no. 41; Marchant's note on Th. 1.51; Böckh, *Kleine Schriften* 6.75.

CHAPTER II

The Place of Thucydides in the Tradition

Thucydides stands apart from lesser historians by virtue of his independence of outlook and his skill in applying the principles of the rhetorical schools to historical writing. It is exceedingly doubtful whether Hellanicus showed any such independence. It is also quite clear that Hellanicus was not greatly influenced by the rhetorical schools, since his style has been characterized as undistinguished and no kind of critical discussion is attributed to him.[1] It is, therefore, unlikely that Thucydides owes any of his stylistic peculiarities or his analytic qualities to Hellanicus. But it is scarcely to be expected that a writer will entirely disregard the traditions of his predecessors or free himself completely from their influence. A search in the text of Thucydides for the features which we have found to be characteristic of Hellanicus is not made in vain.

The object of such a search, however, is not merely to establish a conclusion which most critics would be ready to grant beforehand. The investigation is necessary in order to show that Hellanicus does not stand alone in representing the traditions of Athenian local history in the fifth century. If we find that certain characteristics which are common to the earlier and the later *Atthides* are also to be found in Thucydides, we shall establish a much surer foundation for any remarks we may make about a continuous literary tradition. Thucydides expresses his scorn for the manner in which his predecessors had written history;[2] but it will become apparent that, perhaps unwillingly and unintentionally, he helped to keep alive a tradition which he despised. It will also follow that the later Atthidographers, reactionary in their methods though they appear to be, did not entirely disregard Thucydides. They may have set aside his rhetorical and sophistical characteristics, antagonized, in all probability, by the exaggerated rhetoric and shallow moralizing of Ephorus and Theopompus and others of the Isocratean school.

[1] Hellanic. T.15 (see Chap. 1, note 26).
[2] Cf. his famous remarks in 1.22.

27

But like him, they may still be attempting to produce a "possession for ever"—a truthful account of Attic history without bias or exaggeration. If they fail to produce an enduring work, it is because, in their anxiety not to deviate from the narrow path of historical research, they deliberately shirk certain tasks which we expect a serious historian to undertake; because they attempt to follow strict principles and to conform to a narrow unchanging tradition. Thucydides, fortunately, like Herodotus, did not fetter and confine himself in this way.

This chapter is not concerned with the innovations of Thucydides except in so far as they are adaptations of older methods. Thus, in his selection of a theme, the choice of the Peloponnesian War in particular is a novelty. But there is nothing new in adopting a particular period for a theme; he says himself that his predecessors have dealt either with the Persian Wars or with *Hellenica* previous to the Persian Wars. In selecting a limited period for treatment he throws in his lot with Herodotus, rather than with those who, like Hellanicus, tried to deal with the general history of a particular state. Like Herodotus again, he sets out to give the causes and antecedents of the quarrel between the combatants. Like Herodotus he has digressions: before the scene shifts to Sicily, he gives a brief sketch of Sicilian history and of the Greek settlements there, just as Herodotus seized the opportunity for a digression on Egypt before describing the campaign of Cambyses in that country. In his actual treatment of digressions, however, Thucydides differs greatly from Herodotus. He keeps them within reasonable limits, does not attempt to put down everything he knows, pays little attention to the social customs of peoples. He makes no such ambitious attempt as does Herodotus to trace the previous history of the combatants or to catalogue their earlier disagreements. Indeed, his sketch of the Pentecontaetia is an exception to his rule, and he feels obliged to justify this exception with the excuse that no good account of that period exists.[3] He has no comparable account of the rise and development of Sparta or the Corinthian mercantile empire, of Spartan military traditions or the development of democracy in the various Greek states.

None of these matters would be irrelevant to his main theme and it is easy to think of other topics which he might reasonably have discussed in order to explain more fully the significance of the

3 1.97.

Peloponnesian War. Indeed the absence of such discussions is one of the reasons which have led scholars to believe that his account of the Pentecontaetia is an unfinished sketch. This is not the place to discuss how far the present form of Book I differs from what its author intended it to be. Fourth-century historians could not answer this question any more definitely than we can, and there is nothing to show that they concerned themselves with it at all. Consequently, so far as its influence on later writers is concerned, it makes no difference whether or not Book I was completed in accordance with the original plans of Thucydides, and for the purpose of the present discussion the question is not relevant; it affects the merits of Thucydides as an historian and a literary artist, but it has nothing to do with the part his book played in the development of a literary tradition. The following discussion, therefore, will be based on the book as it stands, in the form in which his literary successors knew it.

In the first book, his digressions are for the most part concerned with imperial developments that might be considered as in part responsible for the war or else with specifically Athenian matters. It is true that his choice of topics is to some extent influenced by the lack of any accurate treatment of these questions. But in his general preference for digressions about Athenian matters he shows an affinity to Hellanicus rather than to Herodotus. His brief account of the ceremony of public burial, his remarks about Athenian religious and social traditions (the Hermae, the rural habits of the Athenians), and his occasional insistence on points of Athenian topography are all in conformity with the traditions of an *Atthis*. These brief digressions or allusions deserve fuller treatment at a later stage in this chapter.

In one of his early chapters Thucydides admits the possibility that the absence of sensational stories and curious legends in his history (τὸ μὴ μυθῶδες αὐτῶν) [4] may disappoint his readers. He means that he will not treat history as though it were myth (an implied criticism of Herodotus) and that he is not primarily concerned with legend at all. Certainly, in comparison with Herodotus and Hellanicus and the Ionian school in general, he devotes very little space to mythological discussion. But it is not entirely lacking, and it is instructive to notice the form which such discussion takes and the occasion for its appearance. He is not concerned with

---

[4] 1.22. For the meaning of μυθῶδες, cf. his use of the word in 1.21.

establishing the detailed truth about early times. Though ready
enough to express his own opinion on questions of policy or recent
political problems, he is not prepared, like Hecataeus, to offer a
dogmatic account of the heroic age.[5] He admits that the history
of very early times cannot be established by research and that the
passage of years has so transformed the tradition that no trust can
be put in it.[6] But this recognition of his helplessness has not
blunted his interest in early history. In justifying his choice of a
theme, his first impulse is to compare the Peloponnesian War not
only with the Persian Wars, but with the wars of the heroic age;
and he ventures to say that, in his opinion, formed only after
careful investigation, the warlike movements of those days were
on a comparatively small scale.[7] These opening chapters are exceedingly interesting and, before the development of archaeology,
formed the basis for scholarly opinion about early Greek history.
It is doubtless true that, in its good sense and lack of prejudice,
this discussion is far superior to any treatment of the same theme
by his Ionian predecessors. But it is in the Ionian tradition none
the less, for it is an attempt by rationalistic methods to extract the
nucleus of truth out of the mass of myth contained in the Homeric
poems and the epic cycle.[8]

The attitude of Thucydides towards Homer is, in a modified
form, the Ionian attitude. He rationalizes the Homeric account
of the Trojan War, maintaining that it was the superior power of
Agamemnon rather than any religious sanctions which enabled
him to rouse the various states against Troy.[9] But with details of
Homeric legend he is not concerned. In speaking of the early
inhabitants of Sicily he is obliged to mention the Cyclopes and
Laestrygonians, but dismisses them with the remark that we must

---

[5] Hecat. F.1. Thucydides in his introductory chapters is much less confident in
expressing his opinions. Cf. 1.1.3: τὰ γὰρ πρὸ αὐτῶν καὶ τὰ ἔτι παλαίτερα σαφῶς μὲν
εὑρεῖν διὰ χρόνου πλῆθος ἀδύνατα ἦν, ἐκ δὲ τεκμηρίων ὧν ἐπὶ μακρότατον σκοποῦντί μοι
πιστεῦσαι ξυμβαίνει, οὐ μεγάλα νομίζω γενέσθαι οὔτε κατὰ τοὺς πολέμους οὔτε ἐς τὰ ἄλλα.

[6] 1.21: ἐκ δὲ τῶν εἰρημένων τεκμηρίων ὅμως τοιαῦτα ἄν τις νομίζων μάλιστα ἃ διῆλθον
οὐχ ἁμαρτάνοι, καὶ οὔτε ὡς ποιηταὶ ὑμνήκασι περὶ αὐτῶν ἐπὶ τὸ μεῖζον κοσμοῦντες μᾶλλον
πιστεύων οὔτε ὡς λογογράφοι συνέθεσαν ἐπὶ τὸ προσαγωγότερον τῇ ἀκροάσει ἢ ἀληθέστερον,
ὄντα ἀνεξέλεγκτα καὶ τὰ πολλὰ ὑπὸ χρόνου αὐτῶν ἀπίστως ἐπὶ τὸ μυθῶδες ἐκνενικηκότα,
ηὑρῆσθαι δὲ ἡγησάμενος ἐκ τῶν ἐπιφανεστάτων σημείων ὡς παλαιὰ εἶναι ἀποχρώντως.

[7] The first twelve chapters of Book I are occupied with this theme.

[8] Note especially his use of εἰκός in 1.4 and 1.10.3, and of εἰκάζειν in 1.9.4. Cf.
also his references to the Homeric poems in 1.3 and 9; 3.104.4.

[9] 1.9.1: Ἀγαμέμνων τέ μοι δοκεῖ τῶν τότε δυνάμει προύχων καὶ οὐ τοσοῦτον τοῖς Τυνδάρεω
ὅρκοις κατειλημμένους τοὺς Ἑλένης μνηστῆρας ἄγων τὸν στόλον ἀγεῖραι.

be satisfied with what the poets have said and everyone is entitled
to his own opinion about them.[10]   But the element of truth in the
story of Scylla and Charybdis is not allowed to pass unnoticed.
When he has occasion to speak of the straits of Messina and the
Syracusan plan to occupy Rhegium as well as Messene, he remarks
that this strait is the traditional site of Charybdis, and that, because
of its difficult currents, it deserves its reputation for danger.[11]

The Ionian writers had found that the surest way to interest
their readers in geography was to connect certain cities and pro-
montories with Homeric legend.   The fragments of Hecataeus'
Periegesis are full of references to Homeric characters and episodes
in mythology.   Familiarity with Homer was a qualification which
every Greek writer expected from his readers.   He expected it far
more confidently than a knowledge of the real geography of the
west.   It is not surprising, then, if Thucydides sometimes falls
back on legend to introduce his readers to unfamiliar lands.   The
contemporary quarrels between Amphilochians and Ambraciots
doubtless seemed unimportant to most Athenians, and it is probable
that many of them never even knew about their existence.   But,
luckily for Thucydides, he can make the situation seem like a
sequel to a familiar legend.   Amphilochus, an Argive, founded
Amphilochian Argos after his return from Troy, dissatisfied with
conditions in his native Argos; then, "many generations later,"
the Ambraciots were admitted to partnership in the settlement and
became hellenized; in time they obtained the upper hand and
expelled the Argive inhabitants; these in their turn asked help
from the Acarnanians and Athenians.[12]   Thus, with a minimum of
technicalities and introductory explanations, Thucydides is ready
to begin his story of the Athenian activities in that country.   Again,
he can readily explain the inability of the Athenians to capture
Oeniadae by reference to the legend of Alcmeon.   He can make his
readers understand the size of the Achelous by reminding them of
Alcmeon: how he could not be free from blood guilt till he found some
land which the sun had not seen when he killed his mother; and

---

[10] 6.2.1: παλαίτατοι μὲν λέγονται ἐν μέρει τινὶ τῆς χώρας Κύκλωπες καὶ Λαιστρυγόνες
οἰκῆσαι, ὧν ἐγὼ οὔτε γένος ἔχω εἰπεῖν οὔτε ὁπόθεν ἐσῆλθον ἢ ὅποι ἀπεχώρησαν· ἀρκείτω
δὲ ὡς ποιηταῖς τε εἴρηται καὶ ὡς ἕκαστός πῃ γιγνώσκει περὶ αὐτῶν.

[11] 4.24.5: καὶ ἔστιν ἡ Χάρυβδις κληθεῖσα τοῦτο ᾗ Ὀδυσσεὺς λέγεται διαπλεῦσαι. διὰ
στενότητα δὲ καὶ ἐκ μεγάλων πελαγῶν, τοῦ τε Τυρσηνικοῦ καὶ τοῦ Σικελικοῦ, ἐσπίπτουσα ἡ
θάλασσα ἐς αὐτὸ καὶ ῥοώδης οὖσα εἰκότως χαλεπὴ ἐνομίσθη.

[12] 2.68.

how he at last discovered the newly formed land made by the deposits of the Achelous and settled on the site of Oeniadae.[13]

It need not be supposed that Thucydides faithfully believes in all these myths or expects his readers to do so. Like Herodotus, he interjects an occasional λέγεται or ὥς φασιν [14] in his accounts of legends, as though to remind his readers that responsibility for belief or disbelief rests with them. On one occasion, indeed, he protests against the incorrect application of myth to contemporary history. He insists that Teres, father of King Sitalces of Thrace, is no relation whatever of Tereus, who married Procne, daughter of the Athenian king Pandion; that the two men came from entirely different parts of the country; and he adds a further argument from probability—that the Odrysians were much too far away for their king to contract an Athenian alliance in early times, whereas Daulis in Phocis was more accessible.[15] Like the Ionian mythographers, he is anxious that legends, whether true or not, should at least be reasonable.

He recognizes, therefore, that it is a mistake to find any connection between the Odrysians and Athenian mythology. But he is anxious to point out any connection that can be established between a semi-barbarian country and old Greek families. Perdiccas, the Macedonian king, is said to be of Argive origin, descended from the Temenidae in Argos.[16] His decision to follow the Argives in withdrawing his support of the Athenians in 418 is supposed to be the result of this ancient Argive connection: "He did not immediately withdraw from the Athenians but contemplated the step, since he saw the Argives doing so; and he traced back his own origin to Argos." [17] Thucydides also thinks it worth while to give the traditional account of how the Macedonian dynasty established itself: how the original settlers drove out the Pierians from Pieria, the Bottiaeans from Bottiaea; how they settled along the Axius and between the Axius and the Strymon, driving back the Eordi and Almopes, and so gradually extended their power.[18] His

[13] 2.102.5.
[14] As in 3.96.1; 6.2.1. Note also the δή in 2.102.5: λέγεται δὲ καὶ 'Αλκμέωνι τῷ 'Αμφιάρεω, ὅτε δὴ ἀλᾶσθαι αὐτὸν μετὰ τὸν φόνον τῆς μητρός, τὸν 'Απόλλω ταύτην τὴν γῆν χρῆσαι οἰκεῖν.
[15] 2.29.3: This introduction of an argument from τὸ εἰκός, after positive proof has been offered, is in the tradition of contemporary Attic oratory.
[16] 2.99.3.
[17] 5.80.2.
[18] 2.99.

account is a *Ktisis* in miniature, an account of the founding of a nation, and it must be remembered that Hellanicus wrote a special work on *Ktiseis*, devoted to the foundation legends of various cities.

Konrat Ziegler, in an interesting article on the excursuses in Thucydides,[19] suggests that these digressions from the main theme are the result of research in the Ionian manner carried out by Thucydides before he formed his project of writing a history of the Peloponnesian War. It is certainly true that these digressions show his affinity to the Ionian tradition and one of them—the so-called *Archaeologia* of Sicily at the beginning of Book VI—is devoted almost entirely to *Ktiseis*. First he enumerates the pre-Greek settlers, beginning with the legendary Cyclopes and Laestrygonians, then the Sicans, the Trojans, the Sicels, and the Phoenicians There is no occasion here to discuss the source of his material;[20] it is the manner of presentation which concerns us. He gives the divergent tradition about the origin of the Sicans, who claimed to be autochthonous though research revealed that they came from the Sicanus in Spain. He uses the appeal to reason (ὡς μὲν εἰκὸς καὶ λέγεται) in support of the story that the Sicels, fleeing from the Opici, crossed the straits of Messina on rafts. He gives the derivation of the name of Italy from the Sicel king Italus, a typical Ionian explanation, perhaps borrowed from Antiochus of Syracuse.[21] In the account of the Greek settlements which follows, care is taken to name the original οἰκισταί and the date of the settlement.

It does not fall within the scope of this chapter to discuss in detail the characteristics of Ionian *Ktiseis* nor the influence exercised on Thucydides by this particular branch of quasi-historical literature. The interest of Thucydides in early settlements is relevant here only because it is one of the points which show his affinity to the Ionian school and the methods of Hellanicus.[22] Apart from the opening chapters of Book VI and his remarks about the origin of Amphilochian Argos, it is worth while to note his mention of the founders (οἰκισταί) of Heracleia in Trachis,[23] his account of the attempts to colonize Amphipolis and their date,[24] and the claim of

---

[19] "Der Ursprung der Exkurse in Thukydides," *RhM* 78 (1929) 58–67.

[20] Antiochus of Syracuse is commonly regarded as a source for Thuc. here. Cf. W. Schmid, *Griech. Literaturgesch.* 1.2.704, where the relevant literature is listed.

[21] Cf. D.H. *Ant. Rom.* 1.35, where the version of Hellanicus is also given.

[22] For Hellanicus' account of the settlement of Sicily, cf. F.79 and 82. For his etymologies of tribal and place names cf. F.13, 14, 38, 71.

[23] 3.92.5.

[24] 4.102.

the people of Scione that the inhabitants of Pellene came originally from the Peloponnese and settled in their new home on their return from Troy.[25]

Far more frequent than these remarks characteristic of *Ktiseis* are those which recall the style of a *Periegesis* and the methods of Hecataeus of Miletus. Such characteristics, as revealed by the fragments of Hecataeus, are a readiness to connect geographical sites with heroes and episodes of mythology; an interest in earlier names of places; a tendency to describe an inland settlement or an island by giving its distance from the coast; and above all a tendency to use certain very brief formulae in giving the information. Numerous sentences can be found in Thucydides, which show him using formulae of description similar to those of Hecataeus. Admittedly, these formulae might be used by any writer of any age and their appearance in itself proves nothing. One is justified in claiming to find the influence of a *Periegesis* here only because there are occasions when this conventional style of description is out of keeping with its surroundings, and the geographical notes seem uncalled for by the narrative or the argument. I have discussed this question at greater length in a special article,[26] and only a small portion of the relevant material can be set forth here.

At the end of Book II Thucydides has a digression on the extent of the Odrysian kingdom and describes the various peoples in it. In this description there are a number of sentences which, to a reader familiar with the fragments of Hecataeus and the later *Periploi*, recall the style and manner of a *Periegesis*. For example, he describes some of the tribes on the boundary of the kingdom as follows: "In the direction of the Triballi, who likewise are independent, the boundary tribes are the Treres and Tilataei; these live to the north of Mount Scombrus and extend in a westerly direction as far as the River Oscius." Then he adds: "And this river has its source in the same mountain range as the Hebrus and the Nestus; it is a desolate and extensive range, bordering on Rhodope."[27] This is an irrelevant note for his purpose, though it would be normal and necessary in a geographical handbook about Thrace.[28]

---

[25] 4.120.1. Cf. 7.57.4.

[26] "Thucydides and the Geographical Tradition," *CQ* 33 (1939) 48–54.

[27] 2.96.4: ῥεῖ δ' οὗτος ἐκ τοῦ ὄρους ὅθενπερ καὶ ὁ Νέστος καὶ ὁ Ἕβρος· ἔστι δὲ ἐρῆμον τὸ ὄρος καὶ μέγα, ἐχόμενον τῆς Ῥοδόπης.

[28] Cf. a sentence from Hecataeus, quoted by Strabo (12.3.22—F.217): ἐπὶ δ' Ἀλαζίᾳ πόλι ποταμὸς Ὀδρύσσης ῥέων διὰ Μυγδονίης πεδίου [ἀπὸ δύσιος] ἐκ τῆς λίμνης τῆς Δασκυ-

Again, he describes Epidamnus from the point of view of the sailor on a coasting vessel (like Hecataeus and the later *Periegetae*): "A city on your right as you enter the Ionian Gulf." [29] He adds a note on the neighbouring barbarian tribe, the Taulantii, who are of no particular interest for the moment, since the tribe plays no further part in the history and is never mentioned again; but a geographical description of Epidamnus would not be strictly complete without mention of its barbarian neighbours. As a final example, his description of Cheimerium may be quoted, whose only importance is that the Corinthian ships anchored there for a single night:

> There is a harbour there, and above the harbour at a little distance from the sea a city called Ephyre, in the Elaeatis district of Thesprotis. Beside the city Lake Acherousia has an outlet into the sea; the River Acheron, after flowing through Thesprotis, debouches into this lake, and the lake takes its name from the river. The River Thyamis, which forms the boundary of Thesprotis and Cestrine, also flows into the sea there, and between these two rivers the promontory of Cheimerium juts out. [30]

The number of geographical remarks which recall the style of a *Periegesis* is far too great for them to be quoted in full here. To anyone not familiar with the fragments the similarity of style suggested by the passages just quoted may not seem particularly remarkable. But there is a striking difference when Thucydides describes some site or region which he has seen with his own eyes, such as the district round Amphipolis. Then the conventional manner at once disappears and there is both accuracy and character in his writing. For example: "And Brasidas, realizing this, also took up a position facing them on Cerdylium; this place belongs to the Argilli and is on high ground on the other side of the river not far from Amphipolis, and everything was visible from it, so that Cleon could not have made a move with his army unnoticed." [31]

A reference to the old name of a district does little to make the geography more intelligible, as when he refers to Orchomenus "formerly called Minyan," [32] but by recalling the legendary associations

λιτῖδος ἐς Ῥύνδακον ἐσβάλλει. ἀπὸ δύσιος is certainly incorrect; see *Early Ionian Historians* 70f.
[29] 1.24.1: πόλις ἐν δεξίᾳ ἐσπλέοντι ἐς τὸν Ἰόνιον κόλπον.
[30] 1.46.4.
[31] 5.6.3. Cf. 5.10.6.
[32] 4.76.3.

of a place it doubtless helped to rouse the reader's interest, without the necessity of a mythological digression. Whatever purpose the device served, it was certainly popular with the logographers, and became afterwards a favourite mannerism of Alexandrian poets.[33]

More specifically characteristic of *Atthides* and reminiscent of Hellanicus' *Atthis* are remarks about the topography of Athens, especially when they concern landmarks or monuments of antiquarian interest. The fragments of Hellanicus do not illustrate his fondness for antiquarianism quite so well as the fragments of the later *Atthides* do for their authors. But there is his aetiological explanation of the names of the Areopagus (F.38) and of Munychia (F.42); his account of the origin of the Phorbanteum (F.40), the *heroon* of Stephanephorus (F.46), and the temple of Artemis Colaenis (F.163). Remarks of this kind would be of special interest to the foreigner visiting Athens, in search of the kind of information that a modern traveller expects to find in his *Baedeker*.

In the same manner, when Thucydides has occasion to speak of the site of the temples on the Acropolis, as evidence that the original settlement was on the hill, he is led on to enumerate the older temples and to describe the history and associations of the spring Enneacrounos, formerly called Callirhoe.[34] He points out that the national sepulchre for those who die in battle is in "the most beautiful suburb of the city." [35] He mentions the Ambraciot spoils "still to be seen in the Athenian temples," [36] and describes the Hermae—"the traditional rectangular stone images, found in great numbers both in the doorways of private houses and in temples." [37] He describes the position of Colonus, when recounting the meeting of the assembly there at the time of the revolution: "it is sacred to Poseidon, about ten stades distant from the city." [38] And mention of the Pnyx prompts him to add that this was the place

---

[33] Cf. L. Pearson, "Apollonius of Rhodes and the Old Geographers," *AJPh* 59 (1938) 443–59.

[34] 2.15.4–6.

[35] 2.34.5: τιθέασιν οὖν ἐς τὸ δημόσιον σῆμα, ὅ ἐστιν ἐπὶ τοῦ καλλίστου προαστείου τῆς πόλεως, καὶ αἰεὶ ἐν αὐτῷ θάπτουσι τοὺς ἐκ τῶν πολέμων, πλὴν τοὺς ἐν Μαραθῶνι· ἐκείνων δὲ διαπρεπῆ τὴν ἀρετὴν κρίναντες αὐτοῦ καὶ τὸν τάφον ἐποίησαν. The view of R. Laqueur, "Forschungen zu Thukydides," *RhM* 86 (1937) 316–57, that remarks of this type made in parenthesis are later notes of the author cannot be accepted.

[36] 3.114.1.

[37] 6.27.1: εἰσὶ δὲ κατὰ τὸ ἐπιχώριον, ἡ τετράγωνος ἐργασία, πολλοὶ καὶ ἐν ἰδίοις προθύροις καὶ ἐν ἱεροῖς.

[38] 8.67.2.

where the assembly usually met.[39]   Finally, for the benefit of more painstaking antiquarians, he mentions the altars dedicated by Peisistratus, son of the tyrant Hippias, with the inscription in "faint characters," and the *stele* on the Acropolis commemorating the brutality of the tyrants, on which appear the names of Hippias' five sons.[40]

Parallel with his interest in historic landmarks and monuments, and equally typical of the Atthid tradition, is his interest in traditional Athenian habits, especially religious usage.   He gives a full description of the traditional ceremonies of state burial, as practised after the first year of the war and in other years.[41]   He is careful, in telling the story of the murder of Hipparchus, to point out that the people carried shields and spears (but not daggers) in the Panathenaic procession.[42]   He explains the failure of Cylon's insurrection by his mistake about "the greatest festival of Zeus": he thought these words described the Olympic festival and not the Diasia, in which "the whole people of the Athenians makes offerings, not ordinary victims, but special sacred offerings peculiar to their local tradition." [43]   He also describes, with considerable attention to religious detail, the method in which the Athenians purified Delos in the sixth year of the war.[44]

But his interest in Athenian traditions is not confined to religious ceremonies.   He traces the rural habits of the Athenians back to the way of life which they followed before the time of Theseus, when the different villages in Attica had their own independent *prytanea*—a way of life which was not substantially changed by the synoecism.   He shows his familiarity with the changes made by Theseus, and speaks as though there could be no doubt even about so ancient an event: "When Theseus became king, he showed himself both a shrewd and strong ruler, and besides other improvements he abolished the separate council chambers and officials in the different towns, uniting them all under the present city

[39] 8.97.1: ἐς τὴν Πύκνα καλουμένην, οὖπερ καὶ ἄλλοτε εἰώθεσαν.

[40] 6.54.7, 55.1.   The remark about the faint characters of the inscription is puzzling to archaeologists, since the stone has been discovered and the lettering is perfectly clear.   Cf. M. N. Tod, *Gk. Historical Inscriptions* p. 11.

[41] 2.34.

[42] 6.58.2: καὶ οἱ μὲν ἀνεχώρησαν οἰόμενοί τι ἐρεῖν αὐτόν, ὁ δὲ τοῖς ἐπικούροις φράσας τὰ ὅπλα ὑπολαβεῖν ἐξελέγετο εὐθὺς οὓς ἐπῃτιᾶτο καὶ εἴ τις εὑρέθη ἐγχειρίδιον ἔχων· μετὰ γὰρ ἀσπίδος καὶ δόρατος εἰώθεσαν τὰς πομπὰς ποιεῖν.   This last remark, however, is regarded as an interpolation by many scholars.

[43] 1.126.6.

[44] 3.104.

of Athens, where he established one council chamber and one *prytaneum;* while they continued to live on their own farms as before, he obliged them to use Athens as the only city, which, as all contributed taxes to it, became a substantial city before Theseus handed it on to his successors; and ever since that day the Athenians have celebrated the *Synoikia* as a national festival in honour of the goddess Athena." [45]   Then, after mentioning some of the evidence for the extent of the old city, he goes on: "And so to a great extent the Athenians lived in independent villages all over Attica, and after the union under Athens, following their traditional ways, most of the people, both in earlier and in more recent times, continued to live in the country where they were born; this custom continued right up to the present war, and consequently their removal to the city was not accomplished without hardship." [46]

The story of Theseus occupies a very prominent place both in the *Atthis* of Hellanicus and in the later *Atthides*, and it is interesting to note that Thucydides gives such a careful, though brief, description of his political changes.   Despite his expressed uncertainty over the facts of earlier times, he has no doubt about the way the Athenians lived "in the reign of Cecrops." [47]   There is no trace of polemic in his writing here, and it may be assumed that the traditions of the synoecism were well established.   On the other hand, he does not attempt to solve the vexed question of the quarrel between the Athenians and the Pelasgians,[48] though he has occasion to mention the so-called Pelasgic or Pelargic wall at the foot of the Acropolis, where people camped out in the time of the war, and he mentions the curse on the place and the well known warning of the Delphic oracle:

$$τὸ \ Πελαργικὸν \ ἀργὸν \ ἄμεινον.^{49}$$

Another episode from more recent Athenian history which he describes carefully and without any hint of ambiguity is the revolution of Cylon.   He tells the story with full appreciation of the power wielded by the archons at the time: "In those days the nine archons managed most of the affairs of the city." [50]   He is frankly

---

[45] 2.15.
[46] 2.16.
[47] 2.15.1: ἐπὶ γὰρ Κέκροπος καὶ τῶν πρώτων βασιλέων ἡ Ἀττικὴ ἐς Θησέα αἰεὶ κατὰ πόλεις ᾠκεῖτο κ.τ.λ.
[48] Hdt. 6.137–39.
[49] 2.17.   Cf. 4.109.4.
[50] 1.126.8.

rather scornful about the sincerity of the Spartans in their request
to drive out "the curse" which rested on the Alcmaeonidae by
expelling Pericles,[51] but appears to have no doubts about the facts
of the story. The tradition of Cylon, like that of the synoecism,
was evidently well-established.

Much more controversial is his account of Harmodius and
Aristogeiton.[52] Here too he is not content merely with pointing
out the error of the current versions. He tells the whole story on
an even more flimsy excuse than those which introduce the accounts
of Theseus and Cylon. He says that the fear of tyranny, revived
by the dictatorial ways of Alcibiades, was originally inspired by the
cruelty of the sons of Peisistratus,[53] but that the Athenians were
shamefully ignorant of the facts about Harmodius and Aristogeiton.
The story, which he proceeds to tell, offers a commentary on the
methods of the Peisistratids: their contribution to Athenian prosperity, their quasi-constitutional rule, coupled with dependence on
a strong bodyguard. It is not strictly relevant to the study of
the Peloponnesian War, but it is a distinct contribution to Athenian
local history.

These three incidents, the synoecism, the revolution of Cylon,
and the story of Harmodius and Aristogeiton doubtless occupied
the attention of Hellanicus. They are three of the very few well-
established landmarks in Athenian local history. Of Codrus, Draco,
Solon, and Cleisthenes, equally prominent figures in Athenian local
history, Thucydides says nothing. His only other contribution to
strictly Athenian, as opposed to imperial history, in the period
previous to the Peloponnesian War, is his remark that the Helleno-
tamiae were first appointed when the tribute was assessed for the
allies, and Athens undertook the leadership against Persia.[54] To
judge by the fragments, it was this kind of historical detail which
interested the later Atthidographers rather than the more stirring
incidents of the Pentecontaetia. It is perhaps a curious coincidence

---

[51] Note the δῆθεν in the opening passage of chap. 127: τοῦτο δὴ τὸ ἄγος οἱ Λακε-
δαιμόνιοι ἐκέλευον ἐλαύνειν δῆθεν τοῖς θεοῖς πρῶτον τιμωροῦντες.

[52] 6.54.1: τὸ γὰρ 'Αριστογείτονος καὶ 'Αρμοδίου τόλμημα δι' ἐρωτικὴν ξυντυχίαν
ἐπεχειρήθη, ἣν ἐγὼ ἐπὶ πλέον διηγησάμενος ἀποφανῶ οὔτε τοὺς ἄλλους οὔτε αὐτοὺς 'Αθηναίους
περὶ τῶν σφετέρων τυράννων οὐδὲ περὶ τοῦ γενομένου ἀκριβὲς οὐδὲν λέγοντας. Cf. 1.20.2:
'Αθηναίων γοῦν τὸ πλῆθος "Ιππαρχον οἴονται ὑφ' 'Αρμοδίου καὶ 'Αριστογείτονος τύραννον
ὄντα ἀποθανεῖν, καὶ οὐκ ἴσασιν ὅτι 'Ιππίας μὲν πρεσβύτατος ὢν ἦρχε τῶν Πεισιστράτου
υἱέων.

[53] 6.53.3.

[54] 1.96.

that Thucydides mentions this detail just before recording his
dissatisfaction with the account of this period in Hellanicus and
beginning his own account of it.

Since Thucydides makes the specific complaint against Hellanicus that he gives insufficient information about dates, it is worth
while to examine in some detail the chronological indications which
he gives himself. For the early period of Athenian history he gives
no such indications, and there are no traces of his reckoning by
generations, except the remark that the celebrated sea fight between
the Phocaeans and Carthaginians took place "many generations
after the Trojan War." [55] On the other hand his account of the
colonization of Sicily abounds in dates. The Sicel invasion of
Sicily is said to be "about three hundred years before any Greeks
settled there," [56] and the dates of the Greek settlements are reckoned
from the time of the pioneer settlement at Naxos by the Chalcidians.
No date is offered for the founding of this colony, no indication
given with what events in the old Greek world it coincided. But
the settlement of Syracuse is definitely dated in the following year,
that of Leontini "in the fifth year after Syracuse," that of Gela in
the forty-fifth, and so on.[57] Here Thucydides is evidently following
a well-established tradition, though a purely Sicilian one, which is
not concerned to show the chronological relations with events in old
Greece. Except for the remarks about Hippias—that he was de-
throned in the fourth year of his tyranny and appeared again at
Marathon nearly twenty years later, when an old man—[58] he offers
no date for any strictly Athenian event prior to the fifth century.
The alliance of Plataea with Athens, ninety-two years before its
fall in 427,[59] is an event more important to Plataea than to Athens,
and it is more likely that he knew its date from a Plataean than
from an Athenian source.   So also the dates of the various attempts
to colonize Amphipolis [60] are likely to come from an Amphipolitan
source.

In the fifth century (apart from the year-by-year dating of
events in the war) a number of events are dated exactly. Thus,
the first congress at Sparta, at which both the Corinthians and

[55] 1.14.1.
[56] 6.2.5.
[57] 6.3.1—4.3.
[58] 6.59.4.
[59] 3.68.5.
[60] 4.102.

Athenians speak, is said to take place "in the fourteenth year since the signing of the thirty years truce after the revolt of Euboea." [61] Again, the invasion of Attica by Pleistoanax is said to have been "fourteen years before the present war." [62] A much more elaborate statement of the date of the first hostile act, the Theban attack on Plataea, is given at the beginning of Book II: "The Thirty Years Peace made after the capture of Euboea remained in force for fourteen years; in the fifteenth, when Chrysis had been priestess in Argos for forty-eight years, when Aenesias was ephor at Sparta and Pythodorus still had four months to run as archon in Athens, in the sixth month after the battle of Potidaea, at the beginning of spring." [63] This is the key date for all the subsequent narrative of Thucydides, which makes it possible to dispense with such detailed statements in the rest of his work. A point is being fixed, by which all events connected with the Peloponnesian War may be dated. The actual crossing of the border into Attica was about eighty days after the attack on Plataea.[64] The Peace of Nicias was signed "when winter was merging into spring, immediately after the City Dionysia, when just ten years and a few days had elapsed since the invasion of Attica for the first time took place and the war started." [65] Again, this peace, made when Pleistolas was ephor at Sparta and Alcaeus archon at Athens, remained nominally in force for six years and ten months.[66] The whole Peloponnesian War, including the period of unstable peace, lasted twenty-seven years and a few days.[67] Thus the chronological framework of the history is complete; without any further necessity for naming ephors or priestesses of Hera in Argos, events can be dated by referring them to the winter, spring, or summer of the third, sixth, or fourteenth year of the war.

Nothing could be simpler than such a scheme, and Thucydides defends its accuracy as compared with the system of dating by archons or other officials and saying that an event took place at

[61] 1.87.6.

[62] 2.21.1.

[63] 2.2.1. According to the MSS. Pythodorus had only two months to run, but Krüger's emendation of δύο to τέσσαρας is generally accepted.

[64] 2.19.

[65] 5.20.1.

[66] 5.25.3.

[67] 5.26.1-3. We are not concerned with the strict accuracy of this dating. For a discussion see J. A. R. Munro, "The End of the Peloponnesian War," CQ 31 (1937) 32–38.

the beginning or in the middle of their term of office.[68]   This latter
method seems to have been followed by Hellanicus in the last
portion of his *Atthis;*[69] it was certainly adopted by Philochorus [70]
and most probably by the other Atthidographers as well.   Hellanicus, moreover, in his *Priestesses of Hera,* seems to have tried to
construct a chronological scheme on a Panhellenic scale, dating
events by these priestesses in Argos and the year of their office.
Presumably it is through respect for this work of Hellanicus that
Thucydides refers to the priestess Chrysis in 2.2, and again in 4.133,
when he describes how the temple was burnt and she fled to Phlius,
her place being taken by Phaeinis; he adds that her term as priestess,
thus abruptly terminated, had covered eight and a half years of
the war.

On the other hand, in his *Archaeologia* of Sicily he differs sharply
from the chronology of Hellanicus at one point.   Hellanicus recorded that the Sicels left Italy "in the third generation before the
Trojan War, during Alcyone's twenty-sixth year as priestess in
Argos"; he recorded two waves of migration, first that of the Elymi
driven out by Oenotrians, and then, four years later, that of the
Ausones, driven out by Iapyges.[71]   Thucydides makes these movements subsequent to the fall of Troy,[72] though without any indication that he is correcting his predecessor.   If he owes any of the
dates that follow to the *Priestesses* of Hellanicus, he does not give
any indication of his debt; and his preference for a purely relative
system of dating suggests very strongly that he distrusts the calculations of Hellanicus and is content with the local tradition.

It is also worth noting that Thucydides dates only one event
by the day of the month: the Argive invasion of Epidaurian territory.   Here he gives not the Attic but the Dorian month, the
twenty-fourth day of the month preceding the sacred Carnean
month;[73] for this date, then, his source of information is doubtless
not Attic at all.

[68] 5.20.
[69] Cf. Chap. 1, p. 14 above.
[70] Cf. Didymus, *In D.* (ed. Diels-Schubart) 1.19: προθεὶς ἄρχοντα Νικόμαχόν φησιν
οὕτως.   7.18: προθεὶς ἄρχοντα Φιλοκλέα Ἀναφλύστιον.   13.44: γέγονε δ' αὕτη κατ'
Ἀπολλόδωρον ἄρχοντα, καθάπερ ἱστορεῖ Φιλόχορος.   Cf. below, Chap. 6, pp. 121–132.
[71] F.79b.   Cf. *Early Ionian Historians* 227–30.
[72] 6.2.3: Ἰλίου δὲ ἁλισκομένου τῶν Τρώων τινὲς . . . ἀφικνοῦνται.   Then the Sicels
come: ἔτη ἐγγὺς τριακόσια πρὶν Ἕλληνας ἐς Σικελίαν ἐλθεῖν.
[73] 5.54.3.

Since Thucydides introduces his sketch of the Pentecontaetia
with the complaint that "Hellanicus dealt with this period briefly
and with too little exactness in the matter of dates," the reader
naturally looks for as much accuracy as possible in the twenty
chapters that follow this declaration.[74] This expectation is not
fulfilled, and the history of the period is consequently full of chronological difficulties; the campaigns of Cimon and the Egyptian expedition provide noteworthy examples of such problems. Although one
may excuse brevity and the omission of material on the ground that
Book I lacks final revision, one would expect even preliminary
notes to contain the essential dates; it seems better, therefore, to
blame the inadequacy of his sources for his shortcomings. The
various chronological problems of the period have been discussed
many times,[75] and all that need be pointed out now is the manner in
which Thucydides indicates the passage of years.

The opening chapter of his narrative of the period begins without
any indication of date: "first" comes the capture of Eion; "then"
the settlement of Scyros; "after this" the revolt of Naxos.[76] After
a brief discussion of the grievances of the Athenian allies, we learn
that the Battle of the Eurymedon took place "after this" and the
revolt of Thasos "later on." [77] So it continues right through the
twenty chapters; the same formulae and others like them recur.[78]
When more precise indications of date occur, they are given in
relation to an event which has not yet been exactly dated: we learn
how long the Thasians resisted the Athenians, how long the helots

[74] 1.98–117.
[75] Among recent discussions may be mentioned W. Wallace, "The Egyptian
Expedition and the Chronology of the Decade 460–450 B.C.," *TAPhA* 67 (1936)
252–60. Beloch, in his narrative of the period, departs from the account of Thucydides
very considerably (*Griech. Gesch.* 2.2.178–216); for a rebuttal of his arguments see
W. Kolbe, "Diodors Wert für die Geschichte der Pentekontaetie," *H* 72 (1937) 241–69.
Cf. also Allen B. West, "Thucydidean Chronology anterior to the Peloponnesian
War," *CPh* 20 (1925) 216–37.
[76] 1.98: πρῶτον μὲν 'Ηιόνα τὴν ἐπὶ Στρυμόνι Μήδων ἐχόντων πολιορκίᾳ εἷλον καὶ
ἠνδραπόδισαν, Κίμωνος τοῦ Μιλτιάδου στρατηγοῦντος. ἔπειτα Σκῦρον τὴν ἐν τῷ Αἰγαίῳ
νῆσον, ἣν ᾤκουν Δόλοπες, ἠνδραπόδισαν καὶ ᾤκισαν αὐτοί. πρὸς δὲ Καρυστίους αὐτοῖς
ἄνευ τῶν ἄλλων Εὐβοέων πόλεμος ἐγένετο, καὶ χρόνῳ ξυνέβησαν καθ' ὁμολογίαν. Ναξίοις
δὲ ἀποστᾶσι μετὰ ταῦτα ἐπολέμησαν καὶ πολιορκίᾳ παρεστήσαντο.
[77] 1.100.1–2: μετὰ ταῦτα . . . χρόνῳ δὲ ὕστερον.
[78] Cf. 1.114–115: μετὰ δὲ ταῦτα οὐ πολλῷ ὕστερον Εὔβοια ἀπέστη ἀπὸ 'Αθηναίων. . . .
καὶ μετὰ τοῦτο οἱ Πελοποννήσιοι τῆς 'Αττικῆς ἐς 'Ελευσῖνα καὶ Θριῶζε ἐσβαλόντες ἐδῄωσαν.
. . . καὶ 'Αθηναῖοι πάλιν ἐς Εὔβοιαν διαβάντες Περικλέους στρατηγοῦντος κατεστρέψαντο
πᾶσαν. . . . ἀναχωρήσαντες δὲ ἀπ' Εὐβοίας οὐ πολλῷ ὕστερον σπονδὰς ἐποιήσαντο πρὸς
Λακεδαιμονίους καὶ τοὺς ξυμμάχους τριακοντούτεις.

maintained themselves on Ithome (though there is a textual difficulty here), and how many years the Egyptian expedition lasted; [79] but we do not learn exactly when the hostilities in each instance started. Relative dating of this sort is fairly common,[80] such as would be perfectly satisfactory if we had a solid foundation on which to build, like the second chapter of Book II which provides the basis for all the dates of the Peloponnesian War.[81] The lack of such a basis for his account of the Pentecontaetia makes one wonder whether Thucydides has really justified his criticism of Hellanicus; and the meagre indications in fragments of later historians make one wonder whether very much more precise information was in fact available.

It remains to consider one more point of resemblance between Thucydides and the *Atthides*. The fragments of Philochorus quoted by Didymus show that this author followed an annalistic system in describing the events of the fourth century. Evidence for the closing portion of Hellanicus' *Atthis* is very scanty, but there is just enough to suggest that he used this method for the period of the Peloponnesian War. Lack of fragments renders it quite impossible to know what episodes, if any, he singled out for special treatment or how complete he was in recording the skeleton of military and political movements. It does seem worth while, however, to note what traces there are in Thucydides of an annalistic style, apart altogether from his system of treating the events of each year separately. His method is to single out certain episodes for detailed treatment, but there are other events which he records in bald and brief sentences, in what may fairly be called the style of the chronicler or annalist.

In describing the opening year of the war, Thucydides devotes most of his attention to the incident at Plataea and its immediate consequences, the preparations on both sides for the invasion of Attica and the invasion itself. The offensive movements of the Athenians are treated in rather summary fashion. But there are three chapters in particular which deserve quotation:

[79] Thasos, 1.101.3. Ithome, 1.103.1: οἱ δ' ἐν 'Ιθώμῃ δεκάτῳ (τετάρτῳ, Krüger) ἔτει, ὡς οὐκέτι ἐδύναντο ἀντέχειν, ξυνέβησαν πρὸς τοὺς Λακεδαιμονίους. Egypt, 1.110.1: οὕτω μὲν τὰ τῶν 'Ελλήνων πράγματα ἐφθάρη ἐξ ἔτη πολεμήσαντα.

[80] E.g. 1.112.1: ὕστερον δὲ διαλιπόντων ἐτῶν τριῶν σπονδαὶ γίγνονται Πελοποννησίοις καὶ 'Αθηναίοις πεντέτεις. 1.115.2: ἕκτῳ δὲ ἔτει Σαμίοις καὶ Μιλησίοις πόλεμος ἐγένετο.

[81] Cf. above, p. 41.

And about this same time the Athenians sent out thirty ships off Locris, with the object also of keeping watch over Euboea; the commander of them was Cleopompus, son of Cleinias.   And making some landings he ravaged some of the districts along the seaboard and captured Thronium and took some of the people as hostages; and at Alope he won a battle over some Locrians who came out to resist him. . . .

And in the same summer at the beginning of a new lunar month, which indeed seems to be the only time when it is possible, the sun was eclipsed after midday, and became full again after it had become crescent-shaped and some stars had been visible. . . .

Atalanta was also fortified by the Athenians as a guard post at the end of this summer, an island off the coast of the Opuntian Locrians which was formerly uninhabited; the object was to prevent pirates, who would sail from Opus and elsewhere in Locris, from making raids on Euboea.[82]

Again, in the second year, an attempt of the Spartans to conquer Zacynthus is described in equally summary fashion:

The Lacedaemonians and their allies during the same summer made an expedition with a hundred ships to Zacynthus, the island which lies opposite Elis; the inhabitants are colonists of the Achaeans in the Peloponnese and were in alliance with the Athenians.   On board the ships were a thousand Lacedaemonian hoplites and Cnemus, a Spartiate, was in command.   Landing on the island, they went plundering over almost all of it.   And when the inhabitants did not submit, they returned home.[83]

Subsequent books contain equally short chapters written in the same bald style, often without any explanation of the objects of a movement or its consequences, though it is from such brief statements in Thucydides that the policy of the Athenians in the Archidamian War must be deduced.   In Books VI and VII, when most of the narrative is taken up with his main theme, the progress of the Sicilian expedition, the parenthetic paragraphs describing other incidents of the war are particularly remarkable.   After a descrip-

---

[82] 2.26: ὑπὸ δὲ τὸν αὐτὸν χρόνον τοῦτον 'Αθηναῖοι τριάκοντα ναῦς ἐξέπεμψαν περὶ τὴν Λοκρίδα καὶ Εὐβοίας ἅμα φυλακήν· ἐστρατήγει δὲ αὐτῶν Κλεόπομπος ὁ Κλεινίου. καὶ ἀποβάσεις ποιησάμενος τῆς τε παραθαλασσίου ἔστιν ἃ ἐδῄωσε καὶ Θρόνιον εἷλεν, ὁμήρους τε ἔλαβεν αὐτῶν, καὶ ἐν 'Αλόπῃ τοὺς βοηθήσαντας Λοκρῶν μάχῃ ἐκράτησεν. . . . 2.28: τοῦ δ' αὐτοῦ θέρους νουμηνίᾳ κατὰ σελήνην, ὥσπερ καὶ μόνον δοκεῖ εἶναι γίγνεσθαι δυνατόν, ὁ ἥλιος ἐξέλιπε μετὰ μεσημβρίαν καὶ πάλιν ἀνεπληρώθη, γενόμενος μηνοειδὴς καὶ ἀστέρων τινῶν ἐκφανέντων. . . . 2.32 : ἐτειχίσθη δὲ καὶ 'Αταλάντη ὑπὸ 'Αθηναίων φρούριον τοῦ θέρους τούτου τελευτῶντος, ἡ ἐπὶ Λοκροῖς τοῖς 'Οπουντίοις νῆσος ἐρήμη πρότερον οὖσα, τοῦ μὴ λῃστὰς ἐκπλέοντας ἐξ 'Οποῦντος καὶ τῆς ἄλλης Λοκρίδος κακουργεῖν τὴν Εὔβοιαν.

[83] 2.66.

tion of the preliminary negotiations between the Athenians and the Egestaeans, before going on to give the speeches of Nicias and Alcibiades in the assembly next year, he gives a brief and entirely formal record of the Spartan invasion of the Argolid and the Argive resistance with Athenian aid; and also of the Athenian movements in Macedonia.[84]   Again in Book VII, where the unity of the narrative is even more complete than in Book VI, there occurs this isolated sentence: "During the same summer, towards its close, the Athenian general Euetion, in conjunction with Perdiccas, also made an expedition against Amphipolis with a large number of Thracians; he did not capture the city, but bringing triremes round into the Strymon besieged it from the river, using Himeraeum as his base."[85]   Amphipolis and the Athenian difficulties in Thrace seem so far removed from and almost irrelevant to the problems of Nicias in Sicily, that most readers would pass this chapter by, scarcely noticing it.   Nothing has been said about Amphipolis or the Thracian situation in the whole of Book VI, no attempt is made to trace developments there since the Peace of Nicias; consequently this isolated record of Euetion's campaign seems almost pointless.

In Book VIII the situation is entirely different.   Here there is scarcely any attempt to concentrate on a single aspect of the war as in Books VI and VII.   The historian records different events all over the Greek world without singling out particular episodes for special treatment, and the consequent lack of continuity in the book renders it less interesting and makes this period of the war far less vivid than the earlier years.   Hence the reader is less likely to be startled by the occurrence of passages in annalistic style such as those quoted in the previous paragraphs.   There is not the same remarkable distinction in style between one chapter and another. The reason for these special characteristics of Book VIII, its lack of speeches and critical passages, though often discussed, has never been satisfactorily established, and the question cannot be taken up here.   In this book, perhaps because the war has now officially started again, Thucydides seems much more concerned to be complete in his account than formerly.   Consequently the annalistic passages become less brief and concise, whereas episodes of the type that previously enjoyed full treatment are dealt with in less detail.

[84] 6.7.
[85] 7.9.

It appears, indeed, that the author is trying to find a compromise between the method of the annalist and that of the selective and critical historian. Whatever the truth may be, the result was disastrous for the future of history writing. Xenophon modelled his style on that of Book VIII rather than on the earlier style, thus preparing the way for a return to the older type of chronicle history.

From the point of view of the present investigation Book VIII presents certain other peculiarities apart from those generally emphasized. It contains no mythological digression at all and none of the characteristic marks of *Ktiseis* which are found scattered through all the other books. It is full of geographical parentheses, remarks characteristic of a *Periegesis*, more so than the earlier books; but such remarks never include mention of an old geographical name or lead to any mythological allusion as is the case elsewhere. The two remarks about the topography of Athens, describing the position of Colonus and the fortifications of the Peiraeus,[86] include no allusion to the legendary associations of the former nor the history of the latter place, such as he might reasonably have added. There is no mention of any date beyond the normal division of activities into years and seasons. In other words, except for the brief passages of geographical explanation, the features which have been found to be characteristic of an *Atthis* are lacking in this book.

In general, however, except for the refusal to date events by archonships, Thucydides has not discarded any of these features. His historical work stands apart from the *Atthides* because his innovations are of such a striking nature as entirely to overshadow his connection with the older tradition. The foregoing discussion, if taken by itself, would be absurdly one-sided and disproportionate as a critical estimate of his work. Indeed, for the student of Thucydides these traditional characteristics are quite unimportant. They are important only for the student of a literary tradition, who is concerned to see whether or not Thucydides has completely broken away from it, and their presence shows that he has not been able to do so. Such loyalty to a tradition cannot be argued as a point either for or against him. The same thing might be said, though in less categorical terms, of Herodotus, whose relation to the Ionian tradition is a much closer one and who makes a less

---

[86] 8.67.2, 90.4–5.

determined and conscious effort to strike out a new line for himself; he cannot be blamed for following a well-established tradition, but he must certainly be praised for having breathed new life into it. Both Herodotus and Thucydides have peculiar individual touches which distinguish their work from the commonplace. Their successors made little or no effort to copy these individual touches; and they substituted nothing in place of them to enliven the bare bones of annalistic narrative, except when they tried to make history conform to the rules of the rhetorical exercise.

CHAPTER III

THE SUCCESSORS OF THUCYDIDES

Just as the real distinction of the work of Thucydides is something quite apart from the traditions of Attic local history, so also the new tendencies of history writing in the fourth century are alien both to the Atthis tradition and to the old spirit of Ionian historiography. The link between history and oratory becomes a much closer one, and the pupils of Isocrates lead the way in indulging a fondness for moral reflection. The fragments of the Atthidographers, on the other hand, show little or no trace of these new tendencies. It is not necessary, therefore, for the purpose of the present discussion, to examine the special characteristics of fourth century authors, any more than it was necessary to discuss the distinguishing features of the history of Thucydides.

At the same time, however, the characteristics which were discussed in the preceding chapters continue to appear in greater or less degree in all the historians of the fourth century. The interest in mythology and Homeric interpretation, in geographical detail and traditions about the founding of cities is still apparent; traces of it can be found in the fragments of Ephorus and Theopompus as well as in Xenophon's *Hellenica* and Aristotle's *Constitution of Athens.* In examining the work of Cleidemus and Androtion we cannot speak of the influence of Ephorus and Theopompus, since all these writers were contemporaries. But it will be worth while first to see what part the familiar traditional elements play in the work of the more immediate successors of Thucydides.

There is one writer belonging to the fifth century, a contemporary rather than a successor of Thucydides, whose work calls for brief discussion—Stesimbrotus of Thasos.[1] It is true that the fragments (which are very few in number) seem to represent him as more typical of the fourth century than of the fifth; but Plutarch and Athenaeus insist that he is a contemporary of Cimon and

---

[1] The fragments are in *FGrH* 2 B, no. 107. See also W. Schmid, *Griech. Literaturgesch.* 1.2.676–78, R. Laqueur, *RE s.v.* "Stesimbrotos," Busolt, *Gr. Gesch.* 3.1.7–31, Wilamowitz, *H* 12 (1877) 361–67.

Pericles,[2] and in the Platonic dialogue *Ion* he is held up as one of the standard fifth century authorities on Homer.[3]   He is, in fact, a precursor of the fourth century writers; like the Atthidographers, he combined discussion of Athenian history with that of religious ritual; and like Ephorus and the Ionians he was interested in Homeric problems.   He is credited with a work *On Themistocles, Thucydides, and Pericles*, to which Plutarch refers for a number of unimportant and untrustworthy anecdotes about the personal life of Cimon, Themistocles, and Pericles.[4]   This type of anecdote plays its part in Herodotus, but is almost completely lacking in Thucydides; he does not even record the gossip about Pericles and Aspasia, so that without Plutarch's help it would be difficult for us to appreciate the *Acharnians*.   In the fourth century the emphasis on scandalous detail was very general; Theopompus, for example, is said to have taken particular delight in "revealing the mysteries of false righteousness and unsuspected villainy."[5]

This gossipy or (as Wilamowitz maintains) venomous work of Stesimbrotus would probably be interesting reading. The fragments of his other work *On Ritual* (Περὶ Τελετῶν) are trifling;[6] but the title is worth noticing; it shows that there is a literary precedent in the fifth century for Philochorus' work *On Sacrifices* (Περὶ Θυσιῶν).

Xenophon's *Hellenica* calls for a closer investigation, since it is the only complete historical work that concerns us in this chapter. Xenophon was anxious, particularly in the first two books, to be loyal to the Thucydidean tradition, and he is consequently sparing with his anecdotes.[7]   By omitting the less creditable stories about

---

[2] Plu. *Cim.* 4.5; Ath. 13.589D—*FGrH* T. 1 & 2.

[3] 530D—T.3: οἶμαι κάλλιστα ἀνθρώπων λέγειν περὶ Ὁμήρου, ὡς οὔτε Μητρόδωρος ὁ Λαμψακηνὸς οὔτε Στησίμβροτος ὁ Θάσιος οὔτε Γλαύκων οὔτε ἄλλος οὐδεὶς τῶν πώποτε γενομένων ἔσχεν εἰπεῖν οὕτω πολλὰς καὶ καλὰς διανοίας περὶ Ὁμήρου ὅσας ἐγώ.

[4] *Them.* 2.3—F.1: καίτοι Στησίμβροτος Ἀναξαγόρου τε διακοῦσαι τὸν Θεμιστοκλέα φησὶ καὶ περὶ Μέλισσον σπουδάσαι τὸν φυσικόν. *Per.* 13.16—F.10b: Στησίμβροτος ὁ Θάσιος δεινὸν ἀσέβημα καὶ μυθῶδες ἐξενεγκεῖν ἐτόλμησεν εἰς τὴν γυναῖκα τοῦ υἱοῦ κατὰ τοῦ Περικλέους. Cf. Ath. 13.589D—F.10a: ἦν δ' οὗτος ἀνὴρ πρὸς ἀφροδίσια πάνυ καταφερής, ὅστις καὶ τῇ τοῦ υἱοῦ γυναικὶ συνῆν, ὡς Στησίμβροτος ὁ Θάσιος ἱστορεῖ, κατὰ τοὺς αὐτοὺς αὐτῷ χρόνους γενόμενος καὶ ἑωρακὼς αὐτόν, ἐν τῷ ἐπιγραφομένῳ Περὶ Θεμιστοκλέους καὶ Θουκυδίδου καὶ Περικλέους.

[5] D.H. *Pomp.* 6.7.

[6] F. 12–20.

[7] Cf. L. Breitenbach, *Xenophon's Hellenika* (ed. 2, 1884) 1.22–23. The controversy about the date of composition of different portions of the *Hellenica* is not relevant here; for a lucid summary see J. Hatzfeld, "Notes sur la composition des Helléniques," *RPh*, 3me. sér., 4 (1930) 113–27, 209–26. Cf. also M. MacLaren, "On the Composition of Xenophon's Hellenica," *AJPh* 55 (1934) 121–39, 249–62.

Lysander he succeeds in producing a more flattering portrait of the man; and indeed he prefers to ignore rather than contradict calumnies against those whom he admires. Again, it is interesting to compare the treatment of Agesilaus in the history with that which he receives in the special essay devoted to him. In the *Hellenica* only those stories are told which the author conceives to be historically important: the quarrel with Leotychides about his accession, his emulation of Agamemnon in sacrificing at Aulis, his matchmaking negotiations with Otys and Spithridates, and so on. In fact he apologizes for recording the sayings of Theramenes after his condemnation, on the ground that they are not really "worthy of notice" (ἀξιόλογα), but points out how they reveal the character of the man, who did not lose his sense of humour even at the point of death.[8] He is also apologetic in recording the remark of the machine-builder in Athens,[9] who, when the party from Piraeus was attacking the supporters of the Thirty in the city, gave orders to bring up "rocks fit for wagons to carry" so as to block the way. It is these rare picturesque touches that redeem the *Hellenica* from complete dullness, but Xenophon feels a sense of guilt in thus enlivening his narrative. His sense of what is historically interesting is certainly different from that of Thucydides, but, being first in the field, he can choose his material instead of parading his cleverness by revealing what others have suppressed or ignored.

Xenophon's intention to be loyal to the Thucydidean tradition is probably one of the reasons why touches reminiscent of Hellanicus and the Ionians are scarcer in the *Hellenica* than in Thucydides. So long as the war lasts he preserves the chronological scheme of Thucydides, recording the events of each year separately and adding brief notices of events which he does not describe in detail.[10] The opening sentences of Book I offer a model of annalistic conciseness. At the end of the first chapter, as it is preserved in the manuscripts, there is another typically annalistic sentence about Hannibal's invasion of Sicily "with 100,000 troops," when "in the course of three months he captured two Greek cities, Selinus and Himera."[11]

---

[8] 2.3.56.

[9] 2.4.27: εἰ δὲ καὶ τοῦτο δεῖ εἰπεῖν τοῦ μηχανοποιοῦ τοῦ ἐν ἄστει.

[10] Cf. L. Breitenbach, *op. cit.* p. 9—"viele, mitunter bis zur Unverständlichkeit, kurze Notizen begegnen, die nicht unwichtige, zum Teil sehr bedeutende Ereignisse betreffen." He cites 24 examples from the first two books, and also points out that in Xen. it is not always clear (as it is in Thucydides) what time of year these events take place (p. 41).

[11] 1.1.37.

It seems to be the capture of these two Greek cities, familiar from earlier events in Sicily, that, to his way of thinking, makes the campaign relevant to the *Hellenica;* as a description of the events of the campaign, the sentence is so absurdly inadequate as to be useless; it is no more than a chronological indication and, indeed, almost certainly an incorrect one; most historians place this campaign a year later.[12]   A similar sentence occurs at the end of the second chapter: "And so this year came to a close, in the course of which also the Medes revolted against the Persian king Darius and returned to their allegiance again." [13]   There are brief allusions in the same style to the later Carthaginian attack on Sicily at the end of chapter 5 and to the rise of Dionysius at the end of chapter 2 in the following book.

The third chapter of Book I opens in the same annalistic style, with mention of the burning of the temple of Athena in Phocaea. But, in the typically Thucydidean sentence which follows in the manuscripts, the indications of date are incorrect and generally regarded as interpolations: "And when the winter came to an end, with Pantacles as ephor and Antigenes archon, at the beginning of spring, twenty-two years of the war having passed, the Athenians sailed in full force to Proconnesus." [14]

The likelihood of interpolation makes it difficult to be quite sure how much Xenophon intended to give in the way of chronological indications.   He is certainly not as conscientious as Thucydides in marking the beginning of each summer and winter season and each new year of the war; on several occasions the passage of time has to be deduced from events in the text; on one occasion the mention of snow in the narrative is the only thing to inform the reader that it is now winter.[15]   The text, as it stands in the manuscripts, contains some allusions to Olympic festivals and an occasional sentence pointing out how many years have elapsed since

---

[12] Beloch, *Gr. Gesch.*[2] 2.2.254–255, in common with other critics regards the chronological errors as due to an interpolator.

[13] 1.2.19: καὶ ὁ ἐνιαυτὸς ἔληγεν οὗτος, ἐν ᾧ καὶ Μῆδοι ἀπὸ Δαρείου τοῦ Περσῶν βασιλέως ἀποστάντες πάλιν προσεχώρησαν αὐτῷ.

[14] 1.3.1.   Hude's text (Teubner ed. 1930) is as follows: τοῦ δ' ἐπιόντος ἔτους ὁ ἐν Φωκαίᾳ νεὼς τῆς Ἀθηνᾶς ἐνεπρήσθη πρηστῆρος ἐμπεσόντος. ἐπεὶ δ' ὁ χειμὼν ἔληγε, [Παντακλέους μὲν ἐφορεύοντος, ἄρχοντος δ' Ἀντιγένους, ἔαρος ἀρχομένου, δυοῖν καὶ εἴκοσιν ἐτῶν τῷ πολέμῳ παρεληλυθότων] οἱ Ἀθηναῖοι ἔπλευσαν εἰς Προκόννησον παντὶ τῷ στρατοπέδῳ.

[15] 2.4.3.   For other examples of this negligence on Xenophon's part see Breitenbach (*op. cit.*) p. 39.

the Peloponnesian War started; but since these indications are incorrect, they are generally regarded as interpolations.[16] Furthermore, in the later books, after the close of the Peloponnesian War, there are no annalistic touches and the chronological scheme is much less strict; and, with no indications of date except of the vaguest kind, the result is confusion.

His failure to find an adequate substitute for the annalistic method is significant; the return of the Atthidographers to the old system is easily understood if other methods proved to be unsatisfactory. To Xenophon's account of the forty years which follow the surrender of Athens we can most aptly apply what Thucydides said of the account of the Pentecontaetia by Hellanicus and say he has written βραχέως τε καὶ τοῖς χρόνοις οὐκ ἀκριβῶς. His brevity is to be blamed, because so much is omitted which deserved to be recorded; and his chronology, when it is not merely inexact, is misleading. In this last respect, if in no other, Xenophon apparently resembled Hellanicus more closely than did Thucydides.

Geographical indications are by no means as frequent as they are in Thucydides. Indeed, in the first book a number of comparatively obscure places are mentioned without any indication of their locality: for example, Thoricum, Pygela, Coressus, Chrysopolis, and the "Thraceward" gate of Byzantium.[17] The reader never learns the site of Alcibiades' castle on the Chersonese.[18] In later books the indications are more generous and there are occasional touches reminiscent of a *Periegesis*, such as the description of the Thracian Chersonese in 3.2.10—a necessary description in order to show both the magnitude and the importance of the task which Dercyllidas undertook in building a wall across the isthmus. Arginusae and Aegospotami, as the sites of important battles, are honoured with brief geographical notes.[19] The distance of Sestos from Abydos and that of Ephesus from Sardis is indicated, though Xenophon might reasonably have credited his readers with this knowledge; and it is pointed out that Calydon at one time belonged to Aetolia.[20] Since Xenophon was with Dercyllidas' army in Asia,

[16] Cf. J. Hatzfeld, *Xenophon, Helléniques* (Budé ed.) 1.153–58; G. E. Underhill, *A Commentary to the Hellenica of Xenophon,* xxxvi–xl.

[17] 1.2–3.

[18] 1.5.17.

[19] 1.6.27; 2.1.21.

[20] 4.8.5; 3.2.11; 4.6.1.

his remarks about Leucophrys probably rest on autopsy and are therefore in a somewhat different class.[21]

His remarks on Athenian topography are confined to Book II; not unnaturally, since this is the only book whose interest centres on affairs in Athens. Here too he assumes a certain amount of knowledge, speaking without explanation of the Hippodamian agora in Peiraeus and the road leading to the temples of Munychian Artemis and Bendis; and he speaks of the κωφὸς λιμήν and the mud at Halae.[22] But he adds such curious scraps of information as that the soothsayer of Thrasybulus was buried at the crossing of the Cephisus.[23]

Of allusions to Athenian mythological or archaeological traditions there is no trace, though he sometimes adds a note on a religious festival; for example, he remarks that Alcibiades returned to Athens from exile on the day when the city was celebrating the *Plynteria* "when the statue of Athena was veiled," and that this was a bad omen, because no Athenian ever started work on any important project on that day; and he tells how members of a family were accustomed to meet together at the Apaturia.[24] But he mentions the burning of "the old temple of Athena in Athens" [25] without any reference to its history or its associations for the Athenians. In fact, the absence of such digressions is one of the distinguishing features of the *Hellenica*. The readiness with which Ephorus and Theopompus, like Herodotus, indulged in them is perhaps an indication that in this matter Xenophon made too little concession to popular taste.

In one respect, however, Xenophon is decidedly less severe and critical: in his recital of omens and portents. He clearly believes that these are supernatural warnings and records their supposed occurrence without comment. Earthquakes, thunderstorms, and the omens of sacrifice are all reported in similar style. Very different is the manner of Thucydides in describing the effect of "certain peals of thunder" (ξυνέβη βροντάς τε ἅμα τινὰς γενέσθαι) on the Athenian fighters at Syracuse: how the inexperienced were

[21] 3.2.19.
[22] 2.4.11; 2.4.31; 2.4.34.
[23] 2.4.19.
[24] 1.4.12; 1.7.8.
[25] 1.6.1.

alarmed, but the experienced merely recalled what season of the year it was and took no notice.[26]

In general, it appears that Xenophon played a small part, much smaller than Thucydides, in keeping alive the traditions of Attic local history. Furthermore, since the *Hellenica* is a complete work (though it may lack final revision), one is impressed by certain characteristics which even a large number of fragments might not reveal. Fragments can never show how much an historian failed to mention, and very rarely can they illustrate national partisanship so clearly as the *Hellenica* shows Xenophon's partiality in favour of Sparta. Fragments enable us to conjecture some of the particular qualities of individual Atthidographers. For example, we can decide with reasonable confidence that Androtion had oligarchic sympathies; but we cannot know with so much certainty that he took little interest in military affairs as we do know that Xenophon's interests were predominantly in soldiering and country life rather than in political or constitutional matters.

In the *Hellenica of Oxyrhynchus* [27] the nature of the evidence available is different. Passing judgment on an historian on the basis of a short more or less continuous extract, unaided by any independent knowledge of his life and personality, is in some ways more difficult than judging him from a collection of direct and indirect quotations. It is possible to learn more about his style than would be possible from a thousand fragmentary quotations, but a single extract does not enable the critic even to conjecture the peculiar interests of the author. Fortunately, a single sentence with its reference to an "eighth year" and "the beginning of summer," [28] shows that the writer used an annalistic system of chronology similar to that used by Thucydides and in the first books of Xenophon, though it does not establish the starting point of the history. There is also one excellent example of geographical description in the style of a *Periegesis* about the course of the River Maeander—though here again there is uncertainty about the text.[29] But it is the discussion of the causes of the Corinthian War [30]

---

[26] 6.70.1. Cf. Xen. *HG* 3.4.15; 4.3.10; 4.7.4–7.

[27] References to the text are to Jacoby's edition in *FGrH* 2 A, no. 66.

[28] 4.1: τὰ μ]ὲν οὖν ἀδρότατα τῶν [. . . . . ἔτε]ι τούτῳ συμβάντων [οὕτως ἐγένετο· ἀρχομένου] δὲ τοῦ θέρους τῇ μὲν [- - - - -] ἔτος ὄγδοον ἐνειστήκει.

[29] 7.3: ἐπειδὴ δὲ διεπορ[εύθησαν ταῦτα κατεβίβ]ασε τοὺς Ἕλληνας εἰς τὴν Φ[ρυγίαν ἕως ἀφίκοντο πρὸς τ]ὸν Μαίανδρον ποταμόν, ὅ[ς τὰς μὲν πηγὰς ἔχει ἀπὸ Κελαι]νῶν, ἢ τῶν ἐν Φρυγίᾳ μεγίστη [πόλις ἐστίν, ἐκδίδωσι δ']εἰς θάλατταν παρὰ Πριήνην κ[αὶ - - - -.

[30] 1.2–3.

which has aroused most interest among critics; this is quite in the style of Thucydides and gives us evidence enough that we are not dealing with a mere chronicler or compiler.

Considerable importance has been attached to the manner of this passage in all discussions about the authorship of the extract. The unbiased treatment of the question, as well as the digression on the nature of the Boeotian federal system,[31] has led some critics to believe that the author is definitely not an Athenian.[32]   A more correct statement would be that, if the author is an Athenian, he certainly is not bound by the same conventions as the authors of Attic local histories and his work deserves the name of *Hellenica* rather than *Atthis*.   On the other hand, the restrained style, with complete absence of rhetorical embellishment, is the chief argument against accepting either Ephorus or Theopompus as author.   The claims of both these writers have had their day;[33] more recently the tendency has been to prefer the claim of some less distinguished historian and Jacoby has argued in favour of an almost entirely unknown candidate, Daemachus of Plataea.[34]   It is necessary to point out, however, that the few remarks which recall the style of an *Atthis* do not furnish evidence enough for claiming the author as an Atthidographer;[35] such characteristics can be found in comparative abundance in the fragments of Ephorus and Theopompus, side by side with features which stamp them as pupils of Isocrates. But it will be better to postpone the discussion of these two historians until after the fragments of the earlier Atthidographers have been examined.

[31] 11.2–4.

[32] Cf. W. Judeich, "Theopomps Hellenika," *RhM* 66 (1911), 94–139.

[33] The case for Ephorus has been well set forth by E. M. Walker, *The Hellenica Oxyrhynchia* (Oxford, 1913); cf. also E. Cavaignac, "Réflexions sur Éphore," in *Mélanges Gustave Glotz* (Paris, 1932) 1.143–61.   The German critics have been more inclined to support Theopompus; cf. Ed. Meyer, *Theopomps Hellenika* (Halle, 1909) and R. Laqueur, *RE s.v.* "Theopompos" (9) 5A 2193–2205.

[34] "Der Verfasser der Hellenika von Oxyrhynchos," *NGG* phil.-hist. Kl., 1924, 13–18.   Cratippus, who was originally suggested as a possibility by Grenfell and Hunt, receives no support any longer.   The latest discussion of the problem is by H. Bloch, "Studies in Historical Literature of the Fourth Century B.C.," *HSPh*, Suppl. 1, Special volume in honour of W. S. Ferguson (1941) 303–40.   Bloch comes to the conclusion that the author is not any writer otherwise known to us and rejects the formerly accepted view that only a well-known work would have been preserved in Oxyrhynchus.

[35] The view of De Sanctis, who thinks Androtion is the author, will be discussed in the section dealing with Androtion, below pp. 85–86.

CHAPTER IV

The Earlier Atthidographers

I. CLEIDEMUS

Aristotle's *Constitution of Athens* contains some information about Athenian constitutional history which is not recorded in the extant text or fragments of any earlier Greek writer. Consequently, the question of his literary sources has challenged the ingenuity of scholars, and, when the fragments of earlier historians have not supplied a clue, speculative argument has been ready to suggest an answer. The earlier Atthidographers have been suspected as the source of some of Aristotle's information; although political pamphlets, written at the time of the Four Hundred and the Thirty, may be responsible for some of his political tendencies,[1] the need has been felt to postulate some more comprehensive work on Attic political and constitutional history as his source. Wilamowitz argued strenuously in favour of an *Atthis* earlier than any of those already known to us except the *Atthis* of Hellanicus.[2] He was inclined to believe that a work of this kind established a fixed and semi-authoritative version of Athenian political history, which was perpetuated by later Atthidographers.

Unluckily, this view, plausible though it may be in itself, is not substantiated by the available evidence. No ancient author gives any hint that an *Atthis* was published in the intervening period between Hellanicus and Cleidemus. An attempt to determine the character of *Atthides* written in the fourth century must therefore begin with the fragments of this latter author, whom Pausanias characterizes as "the earliest of those who wrote on Athenian local history."[3]

---

[1] For discussion see the works cited in Chap. 5, note 43, and also K. von Fritz, "Atthidographers and Exegetae," *TAPhA* 71 (1940) 91–126.

[2] *Aristoteles und Athen* 260–90. See esp. 286.

[3] 10.15.6—Cleidemus fg. 15. Cf. Plu. *Glor. Athen.* 345E, where Cleidemus comes first in a list of Athenian writers who played no part themselves in the history which they recorded. Pausanias does not include Hellanicus in this group of local historians. But he quotes from other works of Hellanicus and, though he never actually mentions his *Atthis*, he presumably knew it and perhaps consulted it. See above Chap. 1, p. 15.

Of the ancient authors who refer to his work some call him Κλείδημος, others Κλειτόδημος; Athenaeus and Plutarch prefer the longer form, while the lexicographers are inconsistent. There is no reason to suppose that there are two different authors of similar name. Since the form Κλείδημος is well attested in Attic inscriptions, whereas the longer form is never found,[4] it seems safe to conclude that he should be called Cleidemus rather than Cleitodemus, and more modern critics usually call him by this name.

Pausanias, after calling him the earliest of the Atthidographers, goes on to cite his authority for a strange portent at the time of the Athenian expedition to Sicily: the descent of a large number of crows on Delphi, where they tore away the gold from the statue of Athena with their beaks, damaging the spear of the goddess and the owls and the palm tree in the statuary group. Since he emphasizes the early date of Cleidemus, Pausanias evidently thought that he had witnessed this incident or at least had been living when it was reported. But a more positive piece of evidence about his date comes from another fragment. In discussing the division of the Athenian people into naucraries in the time of Cleisthenes, Cleidemus remarked that "they called these divisions Naucraries, just as they now call the hundred sections into which the Athenians are divided Symmories."[5] The taxation groups known as symmories were first instituted for the payment of *eisphora* in 377; then by the law of Periander, in 357–56, the responsibility for the trierarchy was transferred to twenty symmories.[6] Cleidemus' reference to a hundred symmories, as opposed to the twenty set up in 357,[7] is taken by some critics as evidence that he wrote before this date and that he is speaking of the symmories formed in 377; in that case the Cleidemus mentioned in an inscription as γραμματεύς in 383–82 [8] may be the historian himself. But, on the other hand, no other authority ever speaks of as many as a hundred symmories at any time; it is quite possible that this number is simply an error; and when the passage is examined more fully later in the chapter, it will appear more probable that Cleidemus was in fact referring

---

[4] J. Kirchner, *Prosop. Attica s.v.* "Kleidemos."

[5] Fg. 8. This difficult passage will be discussed more fully later in the chapter, pp. 67–68 below.

[6] [Dem.] 47.21.   Cf. Busolt-Swoboda, *Griech. Staatskunde* 1202.

[7] Demosthenes (14.19) is careful to distinguish the 100 μέρη which he proposes to establish from the existing 20 symmories.

[8] *IG* 2².1930.

to the twenty symmories set up in 357.[9] If his *Atthis* was not written until after that date, it probably antedates Aristotle's *Constitution of Athens* by not more than thirty years.[10]

There is one other biographical indication about Cleidemus: Tertullian (though the text is uncertain) says that he died from an excess of pride when he received a golden crown for the excellence of his historical work.[11]

Since a work called 'Ἐξηγητικόν is attributed to him (the title suggests a book of authoritative religious interpretations and explanations), it seems likely that, like his successor Philochorus, he held the office of Exegetes. His authorship of the *Exegeticon* has been called into question;[12] but there are several fragments dealing with religious matters, which might equally well come from an *Atthis* or from some special work like the books on ritual and sacrifices attributed to Stesimbrotus and Philochorus.[13] Whether or not he actually wrote a separate work on religious usage is of secondary importance; the fragments at least show that he did concern himself with details of religious practice.

Even if the references to the *Exegeticon* and the *Nostoi* should be rejected, there is no reason for treating the two citations from the *Protogonia* in the same way.[14] The meaning of this title is an

---

[9] Jacoby, *RE s.v.* "Kleidemos" (1), prefers the later date of composition, while Müller (*FHG* 1.lxxxii) and Poland (*RE s.v.* Συμμορία 1162) prefer the earlier. M. Cary, *CAH* 6.74, accepts the existence of 100 symmories in 377, evidently following U. Kahrstedt, *Forschungen zur Gesch. des ausgehenden 5. u. des 4. Jahrhunderts*, 209. Cf. J. H. Lipsius, *RhM* 71 (1916) 172–75. Wilamowitz, *Aristoteles u. Athen* 286 note, wants to refer the passage to an even earlier date—some time between 394 and 380, when attempts were first made to rebuild an Athenian fleet.

[10] For the date of composition of the *Constitution of Athens* see Sandys's edition (ed. 2), intro. xlix.

[11] *Anim.* 52: Nam etsi prae gaudio quis spiritum exhalet, ut Chiron Spartanus, dum victorem Olympiae filium amplectitur; etsi prae gloria, ut Clitodemus dum ob historiarum praestantiam auro coronatur. *Ob historiarum praestantiam* is the emendation of Reifferscheid-Wissowa for the unintelligible MS. reading *ab historicis diu praestantiam*.

[12] There is only one reference to it. One passage in Athenaeus (9.409F—Fg. 20) refers to "Cleidemus in the *Exegeticon*." Another passage (13.609C—Fg. 24) refers to "Cleidemus in the eighth book of the *Nostoi*." The most recent editors of Athenaeus [following Stiehle, *Ph* 8 (1853) 633] have emended Κλείδημος to 'Αντικλείδης in both these passages, since Athenaeus mentions works of Anticleides under these titles elsewhere (4.157F, 9.384D, 11.466C, 473B–C). Jacoby, *RE s.v.* "Kleidemos" (1), though inclined to agree with Stiehle about the *Nostoi*, retains the *Exegeticon* for Cleidemus. K. von Fritz, *TAPhA* 71 (1940) 93, does not mention this controversy; he is convinced that Cleidemus was an Exegetes.

[13] Müller attributes Fg. 19–23 to the *Exegeticon*.

[14] Fg. 17, 18.

unsolved puzzle; since Athenaeus refers to the first book of both
the *Protogonia* and the *Atthis* for what seem to be parallel passages,[15]
it is arguable that he is quoting twice from the same passage and
is mistaken in thinking that these are two separate works;[16] but
since Harpocration refers to the third book of the *Protogonia*,[17] it
is hard to accept Jacoby's view that it is another name for the
first book of the *Atthis*. There are also five fragments which appear
to come from some work of Cleidemus on a scientific subject;[18]
whether genuine or not, they are not relevant to the present discussion, since they contribute nothing to our knowledge of Cleidemus as an historical writer. To the remaining twenty-five relevant fragments in Müller's collection, there should be added a
reference in the papyrus commentary on the *Aetia* of Callimachus,[19]
first published in 1912, and two references in Photius.[20] Since it
seems impossible to establish with certainty the number of separate
books that Cleidemus wrote and the peculiar characteristics of each
one, it will be best to use the evidence that the fragments offer
without regard for the titles which they quote, in order to form an
estimate of his historical work as a whole.

[15] The subject under discussion is cookery and the duties of heralds as cooks.
Ath. first cites Book I of the *Protogonia* (14.660A—Fg. 17): ὅτι δὲ σεμνὸν ἦν ἡ μαγειρικὴ
μαθεῖν ἐστιν ἐκ τῶν Ἀθήνησι Κηρύκων· οἴδε γὰρ μαγείρων καὶ βουτύπων ἐπεῖχον τάξιν, ὥς
φησι Κλείδημος ἐν Πρωτογονίας πρώτῳ. Then a little later on he cites Book I of the
Atthis (14.660D—Fg. 2): ἐν τῷ πρώτῳ τῆς Ἀτθίδος Κλείδημος φῦλον ἀποφαίνει μαγείρων
ἐχόντων δημιουργικὰς τιμάς. See below pp. 63–64.
[16] Cf. 9.410F—Hellanicus F.2: τὸν δὲ τῷ χερνίβῳ ῥάναντα παῖδα διδόντα κατὰ χειρὸς
Ἡρακλεῖ ὕδωρ, ὃν ἀπέκτεινεν ὁ Ἡρακλῆς κονδύλῳ, Ἑλλάνικος μὲν ἐν ταῖς Ἱστορίαις Ἀρχίαν
φησὶ καλεῖσθαι· δι' ὃν καὶ ἐξεχώρησε Καλυδῶνος. ἐν δὲ τῷ δευτέρῳ τῆς Φορωνίδος Χαιρίαν
αὐτὸν ὀνομάζει. Since the *Histories* of Hellanicus are not known as a separate work, it
is probable that Ath. is quoting two different readings of the same passage in the
Phoronis. See Jacoby's note on the fragment and *Early Ionian Historians* 167.
[17] *S.v.* Πνυκί—Fg. 18.
[18] Fg. 26—30. Four of these references are in Theophrastus, the fifth in Aristotle's
Meteorologica.
[19] *PBerol.* 11521. Cf. Wilamowitz, *SB. Berlin. Akad.*, 1912, 1.544–47; R. Pfeiffer,
Callimachi fragmenta, no. 4, lines 13–20: νῦν τοὺς Ἕλληνας Ἰή[ονας] κέ[κλ]ηκεν ἀπὸ τῶν
Ἀθηναίων πάντ[ας κοι]νῶ[ς·] οὗτοι γὰρ πρότερ[ο]ν Ἰάονες ἐκαλοῦν[το· καὶ] Ὅμηρος ἐπὰν
λέγῃ Ἰάονες ἐλκεσίπεπλοι τοὺς Ἀθηναίους λέγει, ποδήρ[εις γὰ]ρ [χιτῶνας ἐ]φόρ[ου]ν κατ'
ἀρχὰς ὃν τρόπον καὶ Πέρσα[ι Σ]ύρ[οι Καρ]χη[δ]όνιοι. ἱστορεῖ δὲ ταῦτα Κλείδ[ημος ἐν]
Ἀτθίδι.
[20] *S.v.* ἀδίκιον (R. Reitzenstein, *Der Anfang des Lexikons des Photios* 31): ἀδίκιον
δέ τινες φασι τὴν ἐπὶ τῷ ἀδικήματι τιθεμένην ζημίαν. καὶ γὰρ Κλείδημος ἐν τῇ πρώτῃ τῶν
Ἀτθίδων οὕτω γράφει· "Νόσου γὰρ τοῖς Αἰγινήταις γιγνομένης καὶ μαντευομένοις προηνέχθη
τὸ ἀδίκημα καὶ κατεγνώσθη ἐπὶ τούτῳ τὸ ἀδίκιον." *S.v.* ἀΐδρυτα (Reitzenstein 47): τὰ
κακὰ, τὰ κατάρατα, ἃ ἄλλοι αὐτοῖς οὐκ ἂν ἱδρύσαιντο. εἴρηνται δὲ καὶ αἱ σεμναὶ θεαὶ
ἀΐδρυται ὑπὸ Κλειδήμου.

These fragments, including the questionable citations from the *Exegeticon* and the *Nostoi*,[21] may fairly be classified as follows: two, and possibly a third, refer to Athenian topography; seven or even eight refer to events of mythical times, while five are concerned with more recent Athenian events, from the time of Peisistratus to the fourth century; the remaining twelve refer to details in the political, religious, and social customs of the Athenians.

None of the fragments, unfortunately, contains any indication of the chronological method used in the *Atthis*. Equally lacking is satisfactory evidence about the length of the work, the number of books it contained, or the proportion of space allotted to different periods. There are allusions to a first, third, and fourth book, and even to a twelfth.[22] Since the twelfth book is cited for its mention of the Ἀγαμεμνόνια φρέατα, wells supposed to have been dug by Agamemnon, and the reference to the third book mentions both the reforms of Cleisthenes and the symmories, it seems futile even to hazard a guess about the arrangement of material in the different books. The references to Themistocles and to the Sicilian expedition contain no mention of a book-number.[23]

Since the evidence about the organization of subject matter is so meagre, it will be better to discuss the topics mentioned in the fragments without any attempt to conjecture the context in which each topic was introduced. The remarks about religious ritual may be taken first. Athenaeus refers to the *Exegeticon* for some technical details in the Athenian ritual of purification and quotes *verbatim* the instructions given in that work: "Dig a ditch to the west of the tomb; then standing beside the ditch look towards the west; pour water into it, reciting these words: 'May there be cleansing for you, for whom it is right and lawful.' Then pour another libation of unguent."[24] Besides giving instructions for

[21] There are also three other fragments of doubtful authenticity; the manuscript reading does not give the name of Cleidemus in any of them, but it has been restored by emendation for Καὶ ὁ Δῆμος ἐν αʹ Ἀτθίδος (Fg. 1), Καὶ ὁ Δῆμος (Fg. 23), Καὶ Δῆμος δέ (Fg. 9). It is possible that Καὶ ὁ Δήμων is the correct reading rather than Κλείδημος. For Demon, see below pp. 89–90.

[22] Fg. 1, 2, 4, 8, 9.

[23] Fg. 13, 15. The arrangement of books in the other works is equally obscure. Book III of the *Protogonia* is cited for a remark about the Pnyx (Fg. 18) and Book VIII of the *Nostoi* for the "return" of Peisistratus (Fg. 24).

[24] Fg. 20—Ath. 9.409F. These instructions apparently come under a heading entitled "Purifications." Ath. quotes Cleidemus for his use of the word ἀπόνιμμα: ἰδίως δὲ καλεῖται παρ' Ἀθηναίοις ἀπόνιμμα ἐπὶ τῶν εἰς τιμὴν τοῖς νεκροῖς γινομένων καὶ ἐπὶ τῶν τοὺς ἐναγεῖς καθαιρόντων, ὡς καὶ Κλείδημος ἐν τῷ ἐπιγραφομένῳ Ἐξηγητικῷ.

the proper performance of religious rites, Cleidemus also offered explanations of the meaning of sacred names. Suidas quotes "the author of the *Exegeticon*" as saying that the Tritopatores were sons of Uranus and Ge, and were called Cottus, Briareus, and Gyges; [25] Suidas gives this reference after quoting various accounts by other Atthidographers of these mysterious divinities.[26] Again, Suidas and Photius quote Cleidemus' explanation of the term Ὕης as an epithet of Dionysus; he connected it with ὕειν, "to rain," "because," he says, "we offer sacrifices to him at the time when the god sends the rain." [27] This connection of Dionysus with rain recalls Hellanicus' etymology of the name of Osiris, with whom Dionysus was often identified.[28]

It appears, however, that Cleidemus, like Herodotus,[29] was careful not to reveal too much of the mystic lore. He explained that the "seventh ox" was the name of a cake sacred to the moon; but Hesychius does not refer to him when explaining the reason for the name: that there were horns on the cake.[30] So also Thucydides, in his note on the *Diasia*, at which festival this cake was offered in sacrifice, says only that the Athenians "make offerings peculiar to themselves"; it is the scholiast who gives the full explanation.[31] Other fragments refer to the name of the festival *Proarcturia* or

προθεὶς γὰρ περὶ ἐναγισμῶν γράφει τάδε· "'Ὀρύξαι βόθυνον πρὸς ἑσπέραν τοῦ σήματος· ἔπειτα παρὰ τὸν βόθυνον πρὸς ἑσπέραν βλέπε, ὕδωρ κατάχεε, λέγων τάδε· 'Ὑμῖν ἀπόνιμμ', οἷς χρὴ καὶ οἷς θέμις·' ἔπειτ' αὖθις μύρον κατάχεε."

[25] Fg. 19.

[26] The passage in Suidas (according to Adler's text) is as follows, *s.v.* Τριτοπάτορες· Δήμων ἐν τῇ Ἀτθίδι φησὶν ἀνέμους εἶναι τοὺς Τριτοπάτορας, Φιλόχορος δὲ τοὺς Τριτοπάτορας πάντων γεγονέναι πρώτους. τὴν μὲν γὰρ γῆν καὶ τὸν ἥλιον, φησίν, ὃν καὶ Ἀπόλλωνα τότε καλεῖν, γονεῖς αὐτῶν ἐπίσταντο οἱ τότε ἄνθρωποι, τοὺς δὲ ἐκ τούτων τρίτους πατέρας. Φανόδημος δὲ ἐν Ϛ' φησὶν ὅτι μόνοι Ἀθηναῖοι θύουσί τε καὶ εὔχονται αὐτοῖς ὑπὲρ γενέσεως παιδῶν, ὅταν γαμεῖν μέλλωσιν. ἐν δὲ τῷ Ὀρφέως Φυσικῷ ὀνομάζεσθαι τοὺς Τριτοπάτορας Ἀμαλκείδην καὶ Πρωτοκλέα καὶ Πρωτοκλέοντα, θυρωροὺς καὶ φύλακας ὄντας τῶν ἀνέμων. ὁ δὲ τὸ Ἐξηγητικὸν ποιήσας Οὐράνου καὶ Γῆς φησιν αὐτοὺς εἶναι, ὀνόματα δὲ αὐτῶν Κόττον, Βριάρεων, καὶ Γύγην. Since "the author of the *Exegeticon*" is quoted side by side with several Atthidographers, there is good reason to suppose that Cleidemus is meant.

[27] Fg. 21—Suid. and Phot. *s.v.* Ὕης· ἐπίθετον Διονύσου, ὡς Κλείδημος, Ἐπειδή, φησὶν, ἐπιτελοῦμεν θυσίας αὐτῷ καθ' ὃν ὁ θεὸς ὕει χρόνον. Cf. also *Etym. Magnum* and Favorinus.

[28] Hellanicus F.176—Plu. *Isid.* 34.364D. Cf. Hdt. 2.42.2.

[29] Cf. esp. 2.3.

[30] Hesychius *s.v.* βοῦς ἕβδομος· μνημονεύει δὲ τοῦ ἑβδόμου βοός. ὅτι δὲ πέμμα ἐστὶ καὶ τῆς Σελήνης ἱερὸν Κλείδημος ἐν Ἀτθίδι φησίν (Fg. 16). *S.v.* ἕβδομος βοῦς· εἶδος πέμματος κέρατα ἔχοντος. Cf. also *s.v.* βοῦς.

[31] 1.126. The scholiast says that the θύματα ἐπιχώρια were πέμματα εἰς ζῴων μορφὰς τετυπωμένα.

Proerosia, "the sacrifices before ploughing," and to the appointment of παράσιτοι in the cult of Heracles.[32]

Besides these fragments about ritual there are some concerned with the sites of temples or other holy places. In the most extensive of these fragments, from the old grammarian Pausanias, the text is exceedingly corrupt (which is unfortunate since it appears that the actual words from the *Atthis* were quoted) and it is not possible to attempt a translation.[33] There is an allusion to Helicon as the old name of "the hill now called Agra" and the "altar of Apollo of Helicon," as well as to the *metroon* in Agrae. Evidently the quotation is from a topographical passage, and since it comes from Book I it is quite possibly an attempt to describe the topography and buildings of early Athens. Another fragment from Book I establishes the site of the *Melanippeion*, the *heroon* of Melanippus, son of Theseus, in the deme of Melite.[34] Again, Harpocration quotes his explanation of the name of the Pnyx, "because of the crowding there."[35] This is a good example of an etymological explanation, like his explanation of Dionysus Ὕης.[36]

Athenaeus refers three times to Cleidemus for his remarks about the Heralds at Athens and their duties as cooks in preparing sacrifices. In a discourse about the high repute in which cooks have been and ought to be held, after a number of quotations from the poets, the speaker adds that the honourable standing of cookery is shown by the activities of the Heralds at Athens: "For these sustained the duties of cooks and butchers, as Cleidemus says in Book I of the *Protogonia*."[37] There follows a comparison (also

---

[32] Fg. 23, and 11.

[33] Fg. 1—Bekker, *Anecdota Graeca* 1.326 (cf. critical note 3.1105). The text as given by Bekker is as follows: Κλείδημος (sic) ἐν πρώτῳ Ἀτθίδος (see note 21 above). "Τὰ μὲν οὖν ἄνω τὰ τοῦ Ἰλισοῦ πρὸς ἀγορὰν Εἰληθυῖα. τῷ δ' ὄχθῳ πάλαι ὄνομα τούτῳ, ὃς νῦν Ἄγρα καλεῖται, Ἑλικών, καὶ ἡ ἐσχάρα τοῦ Ποσειδῶνος τοῦ Ἑλικωνίου ἐπ' ἄκρου." καὶ ἐν τῷ τετάρτῳ· "Εἰς τὸ ἱερὸν τὸ μητρῷον τὸ ἐν Ἄγραις." For suggested emendations see Müller's note. For the topography of Agrae or Helicon and the *eschara* of Poseidon, see W. Judeich, *Topographie von Athen* 45, note 2, 176.

[34] Fg. 4.

[35] *S.v.* Πνυκί—Fg. 18: Κλείδημος δ' ἐν τρίτῳ Πρωτογονίας, "Συνῇεσαν," φησίν, "εἰς τὴν Πνύκα, ὀνομασθεῖσαν διὰ τὸ τὴν συνοίκησιν πυκνουμένην εἶναι."

[36] The corrupt text of Fg. 25 may also hide a topographical note: Hesychius *s.v.* Προοικίαι· παρὰ Κλειτοδήμῳ· ἐν ἴσῳ τῷ δήμῳ. As it stands this means nothing, but a reference to one of the demes may be intended. Müller records various unconvincing explanations and emendations.

[37] Ath. 14.660A: ὅτι δὲ σεμνὸν ἦν ἡ μαγειρικὴ μαθεῖν ἐστιν ἐκ τῶν Ἀθήνησι Κηρύκων· οἴδε γὰρ μαγείρων καὶ βουτύπων ἐπεῖχον τάξιν, ὥς φησιν Κλείδημος ἐν Πρωτογονίας πρώτῳ. Ὅμηρός τε τὸ ῥέξειν ἐπὶ τοῦ θύειν τάσσει, τὸ δὲ θύειν ἐπὶ τοῦ ψαιστὰ μεταδόρπια θυμιᾶν·

due to Cleidemus) with the duties of heralds in Homeric times, which has been partly anticipated in an earlier book of the *Deipnoso-phistae*.[38]   The speaker proceeds to remark on the sacrificial duties of the censors at Rome and then recalls three Homeric passages illustrating the duties of Homeric heralds; then he refers to Cleidemus again: "And in Book I of the *Atthis* Cleidemus mentions a tribe of cooks who have certain official privileges."[39]   Evidently Cleidemus is referring again to the Heralds; the "tribe of cooks" is none other than the so-called "tribe of heralds" (κηρυκικὸν φῦλον).[40] It becomes clear, therefore, that he discussed the duties of the Heralds, including their task of preparing and cooking the sacrificial victims; and, in true antiquarian style, he traced the descent of their office from the custom of Homeric times.

One of the new references in Photius refers to a pestilence among the Aeginetans and the expiation which the oracle ordered them to pay in order to be free from it.[41]   Since the first book of the *Atthis* is cited, the incident must belong to early times and is very probably connected with the quarrel between Athens and Aegina described by Herodotus in 5.82–88.   Herodotus says nothing of a pestilence in Aegina, but his account shows that there was more than one version of what happened.   The final episode in his account gives the reason for the change in the dress habitually worn by Athenian women.   After the destruction of the Athenian forces sent to Aegina, the Athenian women whose husbands had been killed blinded the sole survivor, plunging the brooches from their himatia into his eyes, and as a punishment for this horrible deed were obliged to give up their old Dorian dress: "They changed their dress for a linen chiton, so as to avoid the use of brooches; to tell

καὶ οἱ παλαιοὶ τὸ θύειν δρᾶν ὠνόμαζον.   ἔδρων δ' οἱ Κήρυκες ἄχρι πολλοῦ βουθυτοῦντες, φησί, καὶ σκευάζοντες καὶ μιστύλλοντες, ἔτι δ' οἰνοχοοῦντες.   Κήρυκας δ' αὐτοὺς ἀπὸ τοῦ κρείττονος ὠνόμαζον.   ἀναγέγραπταί τε οὐδαμοῦ μαγείρῳ μισθὸς ἀλλὰ κήρυκι.   Müller gives only the first sentence of this passage (Fg. 17), but it seems certain that Cleidemus is responsible for the appeal to Homer and that the φησί in the third sentence refers to him.

[38] 10.425E—Fg. 3.   Cf. Busolt-Swoboda, *Griech. Staatskunde* 1058, note 6, and the literature cited there.

[39] 14.660D—Fg. 2: ἐν δὲ τῷ πρώτῳ τῆς Ἀτθίδος Κλείδημος φῦλον ἀποφαίνει μαγείρων ἐχόντων δημιουργικὰς τιμάς, οἷς καὶ τὸ πλῆθος ἐνεργεῖν ἔργον ἦν.   The text of this last clause is corrupt.   Müller reports two suggestions of Siebelis, but the simplest emendation would be ἐνείργειν (even though this compound of εἴργειν is apparently not attested except in the epistles of Phalaris).   Cf. *Il.* 2.183–89 and 18.503 (the assembly depicted on the shield of Achilles): κήρυκες δ' ἄρα λαὸν ἐρήτυον.

[40] Cf. Pl. *Plt.* 260D.

[41] See note 20 above.

the truth, however, this form of dress is not originally Ionian, but Carian; the dress worn universally by Greek women in ancient days was that which we now call the Dorian style." [42] The new fragment from the commentary on Callimachus' *Aetia* shows that Cleidemus wrote about the dress worn by Athenian men in early times; [43] it is to be presumed, therefore, that, perhaps in the same context, he also discussed the change in the dress of the women, either accepting or denying its connection with the tale told by Herodotus.

In discussing the remaining fragments we can follow the chronological order of the events to which they refer. There are three rather inconclusive fragments referring to events of very early times. One is apparently an allusion to the settlement of the Pelasgians at Athens and the wall which they built; [44] a sentence quoted by Porphyrogenitus may possibly refer to the migrations of heroic times; [45] and the other fragment mentions Creusa, daughter of Erechtheus. [46]

We have more substantial evidence about his treatment of the tale of Theseus, to which Plutarch twice refers. Like other authors of *Atthides*, Cleidemus strove to add something of his own to the story. Plutarch writes as follows: [47]

> Cleidemus gives a rather peculiar and very complete account of these matters, beginning a great way back. There was, he says, a general Hellenic decree that no trireme should sail from any port with a larger crew than five men, and the only exception was Jason, the commander of the Argo, who sailed about scouring the sea of pirates. Now when Daedalus fled from Crete in a merchant vessel to Athens, Minos, contrary to the decrees, pursued him with his ships of war, and was driven from his course by a tempest to Sicily, where he ended his life. And when Deucalion, his son, who was on hostile terms with the Athenians, sent to them a demand that they deliver up Daedalus to him, and threatened, if they refused, to put

[42] 5.88.1.

[43] See note 19 above.

[44] Fg. 22—Suid. *s.v.* "Απεδα· τὰ ἰσόπεδα. Κλείδημος· "Καὶ ἠπέδιζον τὴν ἀκρόπολιν, περιέβαλλον δὲ ἐννεάπυλον τὸ Πελασγικόν." Cf. also *s.v.* ἠπέδιζον.

[45] Fg. 7—Const. Porphyrogen. *De Them.* 2.2 (*Patr. Gr.* vol. 113): ἀλλὰ καὶ τὴν ὅλην Μακεδονίαν Μακετίαν οἶδεν ὀνομαζομένην Κλείδημος ἐν πρώτοις 'Ατθίδος· "Καὶ ἐξῳκίσθησαν ὑπὲρ τὸν Αἰγιαλὸν ἄνω τῆς καλουμένης Μακετίας."

[46] Fg. 10—Sch. Eur. *Med.* 19: περὶ δὲ τῆς Κρέοντος θυγατρὸς οὐχ ὁμοφωνοῦσι τῷ Εὐριπίδῃ οἱ συγγραφεῖς. Κλειτόδημος μὲν γὰρ Κρέουσάν φησι καλεῖσθαι, γῆμασθαι δὲ Ξούθῳ, 'Αναξικράτης δὲ Γλαύκην. As Müller points out, the scholiast has evidently confused the Corinthian Creusa (Jason's bride in the *Medea*) with the Athenian.

[47] *Thes.* 19—Fg. 5. The translation is by B. Perrin (Loeb. ed.), with a few changes.

to death the boys and girls whom Minos had received from them as hostages, Theseus made him a polite reply, declining to surrender Daedalus, who was his kinsman and cousin, being the son of Merope, the daughter of Erechtheus. But privately he set himself to building a fleet, part of it at home in the township of Thymoetadae, far from the public road, and part of it under the direction of Pittheus in Troezen, wishing his purpose to remain concealed. When his ships were ready he set sail, taking Daedalus and exiles from Crete as his guides, and since none of the Cretans knew of his design, but thought the approaching ships to be friendly, Theseus made himself master of the harbour, disembarked his men, and reached Cnossus before his enemies were aware of his approach. Then joining battle with them at the gate of the Labyrinth, he slew Deucalion and his bodyguard. And since Ariadne was now at the head of affairs, he made a truce with her, received back the young hostages, and established friendship between the Athenians and the Cretans, who took oath never to commit any act of aggression.

Again, in the story of Theseus' battle with the Amazons, Plutarch emphasizes how Cleidemus strove after detailed accuracy (ἐξακριβοῦν τὰ καθ' ἔκαστα βουλόμενος).[48] But it will be best to refrain from further comment until the treatment of Theseus in other *Atthides* has been considered.

There is an interesting fragment referring to the time of the Trojan War. There are various versions of how the Palladion was brought to Athens through the agency of Demophon, which cannot be traced to their ultimate source. But Demophon does not come to the fore as a legendary figure until the fifth century and then only in Attic tradition;[49] in the earlier tradition of the Trojan War Athens played scarcely any part. According to Cleidemus,[50] it was from Agamemnon that Demophon stole the Palladion "when Agamemnon put in at Athens. He slew a large number of his pursuers. And when Agamemnon complained, they submitted to trial before a court of fifty Athenians and fifty Argives, who were called Ephetae because the decision was entrusted to them by the two parties (οὓς Ἐφέτας κληθῆναι διὰ τὸ παρ' ἀμφοτέρων ἐφεθῆναι αὐτοῖς περὶ τῆς κρίσεως)." The origin of the Ephetae at Athens is enshrouded in mystery for us, and so probably it was for the Athenians themselves. Cleidemus traces its origin to Homeric times, just as he did that of the Heralds; in similar style Hellanicus had described

---

[48] *Thes.* 26—Fg. 6.
[49] Cf. Knaack, in *RE s.v.* "Demophon" (2).
[50] Fg. 12—Suid. *s.v.* Ἐπὶ Παλλαδίῳ.

the founding of the court of the Areopagus in the reign of Cranaus and given an etymology of its name.[51]

Parallel with his attempt to show the origin of the Ephetae is his interest in another vexed question: that of the naucraries. Aristotle explains how, under the constitution of Solon, there were forty-eight naucraries, twelve to each of the four tribes, and how the naucrari in charge of these groups were concerned with arranging direct taxes (εἰσφοραί) and authorizing expenditures.[52] Then, in his account of Cleisthenes, he describes the rearrangement of the tribes and the organization of the demes; and he says that Cleisthenes "also appointed Demarchi having the same duties as the earlier Naucrari; because he made the demes take the place of the naucraries."[53] Photius, after referring to these two passages, though in a rather confused manner,[54] adds: "Cleidemus in his third book says that when Cleisthenes set up ten tribes instead of four, there was also a division of the people into fifty sections; and they called these sections Naucraries, just as now they give the name of Symmories to the hundred sections into which the people are divided."[55] The accuracy of this translation is by no means certain, and it is quite possible that the text of Photius needs emendation. But as it stands there are several difficulties in the statement attributed to Cleidemus. The question of the number of the symmories has already been discussed;[56] but the remark that there were fifty naucraries under the new plan of Cleisthenes seems a plain contradiction of the statements of Aristotle.[57]

There are, as will appear later, other occasions when the Atthidographers disagree with Aristotle on matters of constitutional history.[58] The precise nature of Cleidemus' argument on this point

[51] F. 38. Cf. Chap. 1, p. 13 above.
[52] Resp. Ath. 8.3.
[53] 21.5.
[54] Cf. Sandys' note on 21.5.
[55] Fg. 8—Phot. s.v. Ναυκραρία· . . . ὁ Κλείδημος ἐν τῇ τρίτῃ φησὶν ὅτι Κλεισθένους δέκα φυλὰς ποιήσαντος ἀντὶ τῶν τεσσάρων, συνέβη καὶ εἰς πεντήκοντα μέρη διαταγῆναι· αὐτοὺς δὲ ἐκάλουν ναυκραρίας, ὥσπερ νῦν εἰς τὰ ἕκατον μέρη διαιρεθέντας καλοῦσι συμμορίας. This is neither very elegant nor very lucid Greek. The minor changes suggested by Müller and Siebelis do not really effect a cure. It seems likely that a whole clause has been lost, with resulting confusion in the text. See also P. Giles, Eng. Hist. Rev. 7 (1892) 331.
[56] See pp. 58–59 above.
[57] For an attempt to reconcile the two passages see Busolt-Swoboda, Griech. Staatskunde 881.
[58] See below pp. 82–84.

is hidden,[59] because Photius, in trying to be concise, has not made his meaning clear. But it is evident that a comparison between naucraries and symmories is emphasized. Cleidemus would scarcely have thought the comparison worth making unless he believed that the naucraries, like the symmories after 357, were organized for equipping the fleet, each naucrary being responsible for one ship (this is really the strongest proof that he wrote *after* not *before* 357). Our knowledge of the naucraries is far too slight for us to be sure that such a view is correct;[60] but Cleidemus, whatever other reasons he may have had, certainly could not have resisted the etymological argument; the temptation to connect ναυκραρία with ναῦς [61] would be too strong for a man who connected the Ephetae with ἐφιέναι.

The only other reference to the sixth century is from the disputed *Nostoi:* according to the account given there, after Peisistratus had returned from exile with a woman in the guise of Athena riding on his chariot,[62] he gave this woman, Phya, the daughter of Socrates, as wife to his son Hipparchus; "and for Hippias, who succeeded him as tyrant, he obtained the daughter of the polemarch Charmus, a very beautiful woman." [63]

There remain three fragments referring to the fifth century. Once again there appears to be a conflict between Cleidemus and Aristotle. Plutarch [64] says that, according to Aristotle, the council of the Areopagus provided for the manning of ships in 480 by giving eight drachmae for each man that served: "But Cleidemus represents this also as a trick (στρατήγημα) of Themistocles. His story is that when the Athenians came down to the Peiraeus the Gorgon's head was missing from the statue of the goddess; Themistocles accordingly made a pretense of looking for it, and in the course of a thorough search discovered a quantity of money con-

---

[59] What, for example, did he suppose was the relation now between naucrary and deme?

[60] It is accepted by Busolt-Swoboda (*op. cit.* 569, 599, 771, 817–18). For more detailed argumentation and bibliography see H. Hommel, *RE s.v.* "Naukraria."

[61] This etymology is very generally accepted since the article of F. Solmsen, *RhM* 53 (1898) 151–58, and the objection that the Athenians had no war fleet in the days of Solon and Peisistratus is brushed aside. Since, however, according to Aristotle *Resp. Ath.* 8.3, the naucraries were also organized πρὸς τὰς εἰσφοράς (like the earlier symmories), Cleidemus had more than the etymological argument to justify his comparison.

[62] Cf. Hdt. 1.60.4.

[63] Fg. 24—Ath. 13.609C.

[64] *Them.* 10–Fg. 13.

cealed in people's effects. This money was confiscated, and so there was plenty to pay the men who were to embark on the ships." This is a typical anecdote of Themistocles, comparable to others which Plutarch very probably gleaned from one or other of the Atthidographers.

Again, in the *Life of Aristeides*,[65] he quotes Cleidemus for the tradition that the fifty-two Athenians who fell at Plataea all belonged to the Aiantid tribe, and that this tribe was accustomed to offer to the Sphragitides nymphs the sacrifice which Delphi had commanded in thanksgiving for their victory, receiving the necessary funds from the public treasury. Pausanias[66] says that he recounted many omens which should have deterred the Athenians from setting out on their expedition against Sicily, including the descent on Delphi of a great number of crows, who mutilated the statue of Athena. About these two fragments there is nothing special to remark except that, like most of the others, they illustrate his interest in antiquarian details such as might concern an Exegetes. The lack of further fragments referring to later historical incidents renders it quite impossible to know in what manner he dealt with the events of his own lifetime.

In general, then, there is no evidence to show whether or not Cleidemus deserved the name of historian. In so far as he dealt with religious and political institutions, he took pains to explain their origin and point out parallel institutions in earlier times. Though it is not possible to discover how much space he allotted to different periods, it is evident that he gave considerable attention to the early days of Athens and strove to show the part it had played in heroic times. In this last respect the later Atthidographers certainly followed his example.

BIBLIOGRAPHY

C. G. Lenz and C. G. Siebelis, *Phanodemi, Demonis, Clitodemi atque Istri* ᾽Ατθίδων *et reliquorum librorum fragmenta* (Leipzig 1812), 29–48.

C. and T. Müller, *Fragmenta Historicorum Graecorum* (*FHG*) 1.lxxxii, lxxxvi–vii, 359–65; 4.645.

R. Stiehle, "*Zu den Fragmenten der griechischen Historiker,*" *Ph* 8.632–34.

F. Jacoby, *RE s.v.* "Kleidemos" (1).

References to other works will be found in the notes.

[65] Chap. 19—Fg. 14.
[66] 10.15.6—Fg. 15. See above p. 58.

II. PHANODEMUS

There is no need of a lengthy exposition to show the points of similarity between Phanodemus and Cleidemus. Even a hasty reading of the fragments will show how closely their interests corresponded. Phanodemus was interested in sacred antiquities—in the worship of the Tritopatores, which Cleidemus had discussed, and in the festivals of the *Choes* and the *Chalceia*.[1] His interest in *aetia* is also clearly revealed by the fragments; he explained the name of Artemis Colaenis and why the daughters of Erechtheus were called παρθένοι Ὑακινθίδες; and he told the origin of the well-known scolion

$$\text{Ἀδμήτου λόγον, ὦ 'ταῖρε, μαθὼν τοὺς ἀγαθοὺς φίλει.}^2$$

It is evident that he devoted a great deal of space to early times and sought to establish the remote antiquity of various Attic customs; his fourth book is cited for Colaenus, the early Athenian king who put up the shrine of Artemis Colaenis, and the fifth book for the daughters of Erechtheus. He also dealt with questions of Attic topography, such as the site of the *Leokoreion* and the position of Xerxes' throne from which he watched the battle of Salamis.[3] These points of resemblance, which show an adherence to traditional method, need not be emphasized further.

None of the fragments gives any satisfactory clue as to his date. Müller decided that he must be slightly younger than Cleidemus, but old enough to be contradicted by Theopompus; the evidence is a passage in Proclus, in which Theopompus is cited as reversing the view of Callisthenes and Phanodemus about the supposed Athenian origin of the Saïtes.[4] This opinion about his date is borne out by a series of Attic inscriptions, in which a certain Phanodemus, son of Diyllus (Φανόδημος Διύλλου Θυμοιτιάδης), plays a prominent part. The earliest of these inscriptions records a resolution of the Boule in the year 343–2 to honour Phanodemus with a golden crown for the high quality of his speeches in the council; and a further proposal

[1] Fg. 4, 13, 22.

[2] Fg. 2—Sch. Ar. Av. 873: φησὶ δὲ Ἑλλάνικος Κόλαινον Ἑρμοῦ ἀπόγονον ἐκ μαντείου ἱερὸν ἱδρύσασθαι Κολαινίδος Ἀρτέμιδος, καὶ Φανόδημος ἐν τῇ δ'. See above chap. 1, p. 15. Fg. 3—Suid. s.v. Παρθένοι; cf. Apostol. 14 s.v. παρθένοις ἐξ ἐφάμιλλος (*Paroemiographi Gr.*, ed. Leutsch, 605–06). Fg. 9—Sch. Ar. V. 1231.

[3] Fg. 6, 16.

[4] *FHG* 1.lxxxiii. Procl. *in Ti.* 21e (ed. Kroll, 1.97): τοὺς δὲ Ἀθηναίους Καλλισθένης μὲν καὶ Φανόδημος πατέρας τῶν Σαϊτῶν ἱστοροῦσι γενέσθαι. Θεόπομπος δὲ ἀνάπαλιν ἀποίκους αὐτῶν εἶναί φησιν. Müller also refers to Fg. 15, where Athenaeus quotes the name of Phanodemus before that of Philochorus (ἱστόρησαν Φανόδημος καὶ Φιλόχορος).

that the Demos shall honour him in a similar way.[5]   Again, in 332–1
it is proposed to honour him with another crown for his services to
the sanctuary of Amphiaraus at Oropus, since he had made excellent
arrangements, at his own expense, for celebrating the *Penteteris*
and other ceremonies at the temple.[6]   On the same day in the
assembly he himself proposes the gift of a crown to Amphiaraus,
because of the warm welcome given to Athenians by the god.[7]
Then three years later he receives yet another crown, as a member
of the committee chosen by vote of the people to supervise the
contest and other details connected with the festival of Amphiaraus.[8]
Finally, another inscription, which is not dated, mentions him as
one of the ἱεροποιοί sent from Athens to Delphi for the Pythian
festival, whose duty it was to supervise the start of the official
Athenian delegation on its way to Delphi; [9] he was in distinguished
company, with Lycurgus and Demades among his colleagues.

The Phanodemus of these inscriptions is a man prominent both
in political life and religious services to the state.   He has been
honoured with golden crowns, though not for the same reasons as
Cleidemus.[10]   His father, Diyllus, bears a name which is familiar
as the name of the Athenian historian ("by no means an insignificant
one," according to Plutarch) [11] who reported the Athenian people's
gift of ten talents to Herodotus as a reward for a lecture.   Since
the historian Diyllus wrote towards the end of the fourth century,
he cannot be the father of Phanodemus; but there is no reason why
he should not be the son of the Atthidographer, bearing the same
name as his grandfather.   Adolf Wilhelm [12] was the first to suggest
that the Phanodemus of the inscriptions was the Atthidographer
and father of Diyllus, the historian; and his view has been widely
accepted.

There are several fragments which show the interest of Phanodemus in affairs outside Attica; but the indications that he was
not an Athenian are not definite enough to overrule the evidence
of the inscriptions.   He spoke of the clan of prophets known as

[5] *SIG*[3] 227.
[6] *Ibid*. 287.   Philip had restored Oropus to the Athenians in 338.
[7] *IG* 7.4252.
[8] *SIG*[3] 298.
[9] *Ibid*. 296.
[10] See section on Cleidemus, p. 59 above.
[11] *Mal. Hdti.* 862B.
[12] *Anz. Akad. Wien*, phil.-hist. Kl., 1895, 44–45.   Cf. also Dittenberger's notes on
the above inscriptions, and Kirchner, *Prosop. Attica, s.v.* Φανόδημος and Δίυλλος.

γαλεοί in Sicily, and Hesychius in citing his authority appears to link him with Rhinthon as a citizen of Tarentum; most critics, however, think that the text is at fault here.[13]   He also wrote a work called *Iciaca* [14] about the island of Icos, and Müller follows Siebelis in the suggestion that he may be a native of Icos. This is not a very well-known island, and when a fragment of Callimachus was discovered in which the poet described his conversation with an Ician named Theugenes about religious customs of the island,[15] critics very properly recalled the *Iciaca* of Phanodemus, as a possible source of Callimachus' information.[16]   In the poem about Acontius and Cydippe in the *Aetia* Callimachus acknowledges that he learnt the tale from "Xenomedes of old, who enshrined the whole island (of Ceos) in mythological memory." [17]   This Xenomedes, he continues, wrote about the foundings of the different cities of Ceos, its various inhabitants and changes of name.   Perhaps Phanodemus wrote about Icos in much the same fashion.   A work of this kind would not be incompatible with his Athenian citizenship and the activities revealed by the inscriptions.

Besides the *Iciaca* of Phanodemus Harpocration once refers to his *Deliaca;* [18] and in the second book of his *Atthis* there is an explanation of the old name of Delos, Ortygia.[19]

[13] Fg. 23—Hsch. *s.v.* γαλεοί, μάντεις· οὗτοι κατὰ τὴν Σικελίαν ᾤκησαν, καὶ γένος τι, ὥς φησι Φανόδημος καὶ ῾Ρίνθων Ταραντῖνοι.   The emendation Ταραντῖνος for Ταραντῖνοι is simple.   Cf. Müller's note and Christ-Schmid, *Gesch. der Griech. Lit.* (1920 ed.) 2.1.179.

[14] St. Byz. *s.v.* ᾽Ικός· νῆσος τῶν Κυκλάδων προσεχὴς τῇ Εὐβοίᾳ· ὁ νησιώτης ῎Ικιος· ἔγραψε δὲ Φανόδημος ᾽Ικιακά.

[15] *POxy* 11.1362, R. Pfeiffer, *Callimachi fragmenta nuper reperta*, no. 8.

[16] The allusion to ᾽Ορέστειοι Χόες in the opening couplet recalls Phanodemus' discussion of the Choes (Fg. 13); εἰδότες ὡς ἐνέπουσι in the fragmentary second column looks like a reference by Callimachus to his source.   Cf. L. Malten, "Aus den Aitia des Kallimachos," *H* 53 (1918) 171.

[17] R. Pfeiffer, *Callimachi fragmenta*, no. 9, 53–77.   Cf. W. Schmid, *Griech. Literaturgesch.* 1.2.680.

[18] Fg. 26—Harp. *s.v.* ῾Εκάτης νῆσος· Λυκοῦργος κατὰ Μενεσαίχμου.   πρὸ τῆς Δήλου κεῖταί τι νησύδριον, ὅπερ ὑπ᾽ ἐνίων καλεῖται Ψαμμητίχη, ὡς Φανόδημος ἐν α᾽ Δηλιακῶν. Vossius, *De Historicis Graecis* 399 (ed. Westermann 483) wanted to emend the passage and assign the *Deliaca* to Phanodicus rather than Phanodemus.   Cf. R. Laqueur, *RE s.v.* "Phanodikos."

[19] Fg. 1—Ath. 9.392D: περὶ δὲ τῆς γενέσεως αὐτῶν (sc. ὀρτύγων) Φανόδημος ἐν β᾽ ᾽Ατθίδος φησίν, ὡς κατεῖδεν ᾽Ερυσίχθων Δῆλον τὴν νῆσον, τὴν ὑπὸ τῶν ἀρχαίων καλουμένην ᾽Ορτυγίαν παρὰ τὸ τὰς ἀγέλας τῶν ζῴων τούτων φερομένας ἐκ τοῦ πελάγους ἰζάνειν εἰς τὴν νῆσον, διὰ τὸ εὔορμον εἶναι. . . . The text is uncertain and the quotation is evidently unfinished.   Another foreign reference in Book VII is cited by Ath. 3.114C—Fg. 5, to the Egyptian bread called κύλλαστις.   Cf. Hdt. 2.77.4.

There are some fragments, on the other hand, which seem to reveal an exaggerated Athenian patriotism. In contrast to the Herodotean manner of seeking an Egyptian origin for Greek things, he insisted that the Saïtes were descended from the Athenians; [20] and he subscribed to the view that Teucer came originally from Athens.[21] These statements seem to show that he was anxious to stress the antiquity of the Athenian settlement in Attica. So also, his statement that Persephone was carried off by Pluto, not from Sicily, but from Attica,[22] shows his desire to make Attica play a more prominent part in legendary times. A rather cruder form of national pride is revealed in his assertion that the Persians at the battle of the Eurymedon had six hundred ships; Ephorus gave them only three hundred and fifty.[23] Plutarch also refers to him for another story to Cimon's credit: that when he was at the point of death before the city of Citium, he ordered that his death be concealed from the men; and that it was not discovered either by friend or foe, with the result that the Greek allied force was able to withdraw safely, "thanks to the generalship of Cimon, who, as Phanodemus says, had been dead for thirty days." [24]

Müller records twenty-six fragments of Phanodemus, as compared with thirty for Cleidemus, and there is one probable addition to his collection. In a Paris manuscript of unknown authorship, which contains explanations of proverbial sayings, there is an allusion to the Atthidographer, if his name has been correctly restored instead of Πάνδημος.[25] Owing to the corruptness of the text, it is not certain exactly what remark is attributed to him, but it has

[20] Fg. 7. See note 4 above.

[21] Fg. 8—D.H. *Ant. Rom.* 1.61: τοῦτον δὲ (sc. Τεῦκρον) ἄλλοι τε πολλοὶ καὶ Φανόδημος, ὁ τὴν 'Αττικὴν γράψας ἀρχαιολογίαν, ἐκ τῆς 'Αττικῆς μετοικῆσαί φησιν εἰς τὴν 'Ασίαν, δήμου Ξυπεταίας ἄρχοντα.

[22] Fg. 20—Sch. Hes. *Th.* 913: ἡρπάσθαι δὲ αὐτὴν φασιν οἱ μὲν ἐκ Σικελίας, Βακχυλίδης δὲ ἐκ Κρήτης, 'Ορφεὺς ἐκ τῶν περὶ τὸν ὠκεανὸν τόπων, Φανόδημος δὲ ἀπὸ τῆς 'Αττικῆς, Δημάδης δὲ ἐν νάπαις.

[23] Fg. 17—Plu. *Cim.* 12.

[24] Fg. 18—Plu. *Cim.* 19.

[25] L. Cohn, "Zu den Paroemiographen," *Bresl. philol. Abhand.* 2.2.71, gives the text as follows, as part of a note: τὰς ἐν "Αιδου τριακάδας· καὶ ἀφιδρύματα 'Εκάτης πρὸς ταῖς τριόδοις ἐστὶ καὶ τὰ νεκύσια τριακάδι ἄγεται. τὰ γὰρ νεώματα οὐκ ἀρχαῖα, ὡς Πάνδημος. λεχθείη δ᾽ ἂν ἡ παροιμία ἐπὶ τῶν περιέργων καὶ τὰ ἀποκεκρυμμένα ζητούντων γινώσκειν. Wilamowitz, *H* 34 (1899) 208–09, first suggested Φανόδημος for Πάνδημος. Instead of the unintelligible τὰ γὰρ νεώματα R. Wünsch, *Jahrb. class. Phil.*, Supp. 27 (1902) 119–21, proposed τεσσεράκοστα or τεσσερακοσταῖα. According to this conjectured reading, Phanodemus pointed out that sacrifices in honour of the dead on the fortieth day were a modern innovation, and originally all sacrifices to Hecate were associated with the number three.

something to do with the days that are specially consecrated to Hecate, to whom everything connected with the number three is sacred. A statement on such a subject would be appropriate to Phanodemus, since he was interested in the traditions of sacrifices and festivals.

Two of the fragments which relate to religious traditions are worth recording in detail. Athenaeus gives his account of the origin of the Choes festival:[26] Demophon, as king of Athens, wished to offer hospitality to Orestes, but, since the stain of blood was still upon the stranger, he could not admit him to any temples nor allow him to take part in public sacrifice; accordingly he gave orders for the temples to be closed and cups of wine to be set in front of each member of the party, with a prize of a cake for the one who was first to drain his cup; he also announced that, when they had finished drinking, they were not to take the garlands which they were wearing to the temples (since they could not come under the same roof as Orestes), but each was to crown his own cup with his garland, and the priestess was to take the garlands to the shrine at Limnae and complete the ceremony of sacrifice in the temple; and henceforth the festival was called The Cups (*Choes*). This is evidently the legend to which Callimachus refers in the *Aetia:* at the house in Egypt, where he met the Ician Theugenes, the "Orestean Choes" were duly celebrated each year.[27]

Athenaeus also records his account of how Dionysus received his title Limnaeus ("in the Marshes"):[28] the Athenians used to come to the sanctuary in the marshes bringing the sweet new wine (γλεῦκος), take it out of the wine jars and mix it for the god and then take some for themselves; and Dionysus was called Limnaeus because this was the first occasion when the new wine was drunk mixed with water; and the springs of water were called nymphs and nurses of Dionysus, because the mixture with the water "increased the wine." These explanations seem to be peculiar to Phanodemus and show the type of antiquarianism that interested Callimachus and his circle.

One other fragment is of special interest. Phanodemus pointed out that the festival of the Chalceia, celebrated by smiths and other

[26] 10.437C—Fg. 13.

[27]   ἠὼς οὔτε πιθοιγὶς ἐλάνθανεν οὐδ' ὅτε δούλοις
        ἦμαρ Ὀρέστειοι λευκὸν ἄγουσι Χόες.

Cf. note 16 above.

[28] 11.465A—Fg. 14.

craftsmen in Athens, was dedicated to Hephaestus, not to Athena.[29] This was a matter of personal concern to the historian. The earliest inscription which records the voting of honours to him in the year 343–2 also records the dedication by the Boule of a statue to Hephaestus and Athena Hephaestia, and certain arrangements for this dedication are proposed by Phanodemus himself.[30]

The fragments thus illustrate admirably his interest in religious *aetia*. There are, on the other hand, only two fragments which refer to political institutions. He connected the origin of the court of the Ephetae with the bringing of the Palladion to Athens, though differing in detail from Cleidemus; according to his account, certain Argives on the way home from Troy were killed on landing at Phalerum by some Athenians who did not recognize them; and when Acamas found out what had happened and the Palladion was discovered, the court ἐπὶ Παλλαδίῳ was established in obedience to an oracle.[31] The history of the Areopagus also claimed his attention; Athenaeus cites both Phanodemus and Philochorus for its old function as *censor morum*, how it used to summon and punish spendthrifts and people who had "no visible means of support." [32]

The fragments do not indicate that he had much to say about Athenian customs in his own time. Suidas cites him as an authority for the practice of the Athenians in sacrificing to the Tritopatores before marriage, when they prayed for the birth of children.[33] He also found occasion to point out that the boxes or wallets in which envoys on a sacred errand carried their provisions were called ἀχάναι;[34] and he spoke of a conjuror, who gave the illusion of spurting alternate streams of wine and milk from his mouth, though the liquid was concealed in bladders underneath his clothes and shot upwards as he squeezed them.[35]

---

[29] Fg. 22—Harp. *s.v.* Χαλκεῖα.

[30] *SIG*³ 227. Cf. R. Laqueur, *RE s.v.* "Phanodemos." Two other fragments relating to religious questions must be mentioned. Phanodemus identified the mysterious goddess Daeira with Aphrodite (Fg. 21); and according to his account Artemis substituted a bear, not a stag, for Iphigeneia, when she was on the point of being sacrificed at Aulis (Fg. 10).

[31] Fg. 12—Suid. *s.v.* ἐπὶ Παλλαδίῳ.

[32] Ath. 4.168A—Fg. 15.

[33] Fg. 4—Suid. *s.v.* Τριτοπάτορες.

[34] Fg. 25—Hsch. *s.v.* Ἀχάνας.

[35] Fg. 19—Ath. 1.20A: Διοπείθης δὲ ὁ Λοκρός, ὥς φησι Φανόδημος, παραγενόμενος εἰς Θήβας καὶ ὑποζωννύμενος οἴνου κύστεις μεστὰς καὶ γάλακτος καὶ ταύτας ἀποθλίβων ἀνιμᾶν ἔλεγεν ἐκ τοῦ στόματος. The context is concerned with θαυματοποιοί.

Unluckily there is no indication at all how much space he devoted to historical narrative.   Plutarch cites him on three occasions in his *Lives* of Themistocles and Cimon: for the throne of Xerxes from which he viewed the battle of Salamis, for the battle of the Eurymedon, and for Cimon's order to have his death kept secret;[36] but apart from these three references there is no evidence whatever about his methods or value as a chronicler of events.

BIBLIOGRAPHY

C. G. Lenz and C. G. Siebelis, *Phanodemi, Demonis, Clitodemi atque Istri* 'Ατθίδων *et reliquorum librorum fragmenta* 3–14.
C. and T. Müller, *FHG* 1.lxxiii, lxxxvii, 366–70.
R. Laqueur, *RE s.v.* "Phanodemos."
References to other works will be found in the notes.

III.   ANDROTION

Androtion has received far more notice from scholars than Cleidemus or Phanodemus and fragments from his *Atthis* are more numerous.   These fragments—fifty-nine in Müller's collection and a few others of more recent discovery[1]—reveal his interest in the traditional subjects and show that he received his share of attention from scholiasts and lexicographers.

But before the fragments themselves are discussed it is necessary to decide whether or not he is the same person as the orator whom Demosthenes attacked in his speech *Against Androtion*.   The earlier critics contented themselves with dogmatic statement on this question.   Müller speaks of his predecessors who, "with no arguments to support them,"[2] maintained that the orator and the historian were the same man; and he follows Siebelis in remarking bluntly that the author of the *Atthis* must not be confused either with the orator or with a writer on agriculture mentioned by Theophrastus, Varro, and Columella.[3]   In more recent times, however, scholars have been inclined to take it for granted that the orator and the

---

[36] *Them.* 13, *Cim.* 12, 19—Fg. 16, 17, 18.

[1] For references see the bibliography at the end of this section.
[2] Nullis nisi argumentis (*FHG* 1.lxxxiii).
[3] Thphr. *HP* 2.7.2–3; *CP* 3.10.4; Varro *Rust.* 1.1.9: De reliquis, quorum quae fuerit patria non accepi, sunt Androtion, Aeschrion, etc.   Colum. 1.1.10: Et alii tamen obscuriores, quorum patrias non accepimus, aliquod stipendium nostro studio contulerunt.   Hi sunt Androtion, Aeschrion, etc.   This writer on agriculture is quite clearly a different person from the Athenian Androtion.

historian are the same man.[4]   The evidence for this identification, which is derived in part from inscriptions, was, it must be admitted, mostly unknown to Müller.   He points out, as though it were a decisive point in his favour, that Suidas, though he has no special article on Androtion ἱστορικός, does not say in his article on the orator that he was also an historian.[5]

The anonymous *Life of Isocrates* mentions among the pupils of Isocrates "Androtion, the author of an *Atthis*, who was prosecuted by Demosthenes."[6]   This statement of the biographer was not known to Müller,[7] but any argument for the identification of the orator and the historian must start from it; taken by itself it would not be decisive evidence, but it is borne out by other testimony. Plutarch[8] speaks of men whose literary work was carried out in exile: Thucydides in Thrace, Xenophon at Scillus, Philistus in Epirus, Timaeus in Athens, Androtion the Athenian in Megara, and Bacchylides the poet in the Peloponnese.   One may suppose that Plutarch regards Androtion as an historian (since he calls Bacchylides "the poet" to distinguish him from the historians), and, since the other historians are named in chronological order, that Androtion is not older than Timaeus.   Exile from Athens naturally suggests some degree of political activity and prominence, such as, indeed, the orator Androtion attained: he was a member of the Boule and went as ambassador to Mausolus;[9] he was prosecuted in 354–3[10] in a γραφὴ παρανόμων for proposing that the Boule be crowned although it had not built ships, and he was under suspicion of appropriating sacred property.   This same Androtion (the son of Andron) was honoured by the people of Arcesine in Amorgos for his services to that city (probably about 357–55); and it is

[4] E.g. E. Schwartz, *RE s.v.* "Androtion," and Dittenberger, *SIG*[3] 1.193: Eundem Atthida scripsisse notum est.

[5] Sed gravissimum est, quod neque Suidas neque Schol. Hermogen., qui de A. rhetoris vita agunt et qua in re excelluerit tradunt, eum historicum fuisse dicunt. Suidas writes as follows: Ἀνδροτίων, Ἄνδρωνος, Ἀθηναῖος, ῥήτωρ καὶ δημαγωγός, μαθητὴς Ἰσοκράτους.

[6] Lines 103–05 (Isocrates, Budé ed. p. xxxvi): Ἀνδροτίωνα τὸν τὴν Ἀτθίδα γράψαντα καθ᾽ οὗ καὶ ὁ Δημοσθένης ἔγραψε.

[7] First quoted by Stiehle, *Ph* 8 (1853) 634–35.

[8] *Exil.* 605C.

[9] Dem. 24.12.

[10] This is the date preferred by Schwartz, *RE* 1.2174.   F. Kahle, *De Demosthenis orat. Androtioneae, Timocrateae, Aristocrateae temporibus* (Diss. Göttingen, 1909), prefers the previous year as given by D.H. *Amm.* 4.   See also W. Jaeger, *Demosthenes* (Eng. translation, Berkeley, 1938) 220.

interesting to note that the three cities of Amorgos received mention in the *Atthis*.[11] Another inscription records that he proposed a decree in honour of the two sons of Leucon, rulers of the Kingdom of Bosporus, because of their promise to "take charge of the export of grain"; this was in 347–6.[12] And another record of a vote by the people mentions an Androtion (the father's name is lost) as an ἐπιστάτης.[13]

Androtion the orator is supposed to have been a pupil of Isocrates, and his father Andron sought the company of sophists.[14] It is true that the *Atthis* of Androtion (like other *Atthides*) appears to have no rhetorical tendencies and Müller thinks it unlikely that a member of the Isocratean school would go over to the opposite camp and write an *Atthis*. But to this objection one could reply that the study of antiquities might well appeal to an orator in exile, when the writing of speeches no longer had any purpose. The second inscription cited above is probable evidence that the orator did not suffer exile till after 347; a passage from Didymus corresponds very well with this indication, since it shows not only that the *Atthis* of Androtion mentioned events as late as 344 but also (if the text is correctly restored) that its author actually took part in a debate at Athens in that year.[15] Of the fragments known to Müller none mentioned any event later than the Corinthian War.[16]

Finally there is a fragment of Philochorus which should be quoted. Harpocration, in his note on πομπεῖα, the sacred utensils used in processions at Athens, refers to the charge of misappropriating sacred utensils made by Demosthenes in his speech against Androtion and adds: "The Athenians, so Philochorus says, pre-

---

[11] *SIG*³ 1.93: ἔδοξεν τῇ βουλῇ καὶ τῷ δήμῳ τῶν Ἀρκεσινέων· ἐπειδὴ Ἀνδροτίων ἀνὴρ ἀγαθὸς γέγονε περὶ τὸν δῆμον τὸν Ἀρκεσινέων . . . στεφανῶσαι Ἀνδροτίωνα Ἄνδρωνος Ἀθηναῖον χρυσῷ στεφάνῳ, κτλ. His presence in Amorgos is probably to be dated during the Social War. Cf. also Fg. 19—St. Byz. *s.v.* Ἀρκεσίνη, μία τῶν τριῶν πόλεων τῶν ἐν Ἀμόργῳ τῇ νήσῳ. ἦσαν γὰρ Μελανία, Μινώα, Ἀρκεσίνη . . . τὸ ἐθνικὸν Ἀρκεσινεύς. Ἀνδροτίων ἕκτῃ Ἀτθίδος· ''Ἀμοργίοις, Μινωΐταις, Ἀρκεσινεῦσιν.'' Müller comments: Vox Ἀμοργίοις merito corrupta esse videtur Siebeli. Requiritur gentile Μελανίας urbis.
[12] *SIG*³ 1.206.
[13] *IG* 2.1².61.
[14] Pl. *Prt.* 315c, *Grg.* 487c.
[15] *In D.*, col. 8, 8–15: ἐπὶ ἄρχοντος Λυκίσκου (344–3) βασιλέως πρέσβεις συμπροσήκαντο οἱ Ἀθηναῖοι, ἀλλ' ὑπεροπτικώτερον ἢ ἐχρῆν διελέχθησαν αὐτοῖς. εἰρηνεύσειν γὰρ πρὸς Ἀρταξέρξην, ἐὰν μὴ ἐπὶ τὰς Ἑλληνίδας ἴῃ πόλεις. ἀφηγοῦνται ταῦτα Ἀνδροτίων, ὃς καὶ τότ' εἶπε, καὶ Ἀναξιμένης. The restoration ὃς καὶ τ[ότ' εἶπε] is by no means certain.
[16] Fg. 50 refers to the Spartan victory near Corinth.

viously used the πομπεῖα which had been obtained out of the property of the Thirty; later on, he says, Androtion provided another set." [17] If Androtion, the orator, who was suspected of stealing sacred property, wrote an *Atthis*, a golden opportunity presented itself to discuss the whole history of πομπεῖα in his book, concluding with a justification of his own actions. Hence it is arguable that Philochorus is quoting, not from his knowledge of the suit against Androtion, but from the *Atthis* of Androtion.

There is, therefore, enough evidence to make the identification of the two men almost certain. Further evidence will appear as the fragments are discussed more in detail.

The *Atthis* is the only work attributed to Androtion and it evidently contained at least eight books.[18] But, as usually happens, we do not know precisely how he arranged his material and what ground was covered in the different books. Harpocration and Stephanus of Byzantium frequently refer to a particular book of his for the name of a city or island, in the same manner in which they cite Hecataeus. Sometimes a reference of this kind suggests a definite incident—for example, Arginusae was evidently mentioned in Book IV on the occasion of the battle; [19] but usually the only inference we can draw from citations of this sort is that the author did not confine himself to the purely domestic history of Athens.

The only reference to the first book is for the establishment of the Panathenaic festival by Erichthonius (Fg. 1). In Book II we find already a reference to the Peisistratid Hipparchus as the first victim of ostracism in 488 (Fg. 5). Book III dealt with the revolution of the Thirty (Fg. 10, 11); and Stephanus cites the book for its mention of Panactum (Fg. 8), which plays a prominent part in the closing years of the Archidamian War. The battle of Arginusae was evidently described in Book IV (Fg. 14). Androtion seems to have wasted little time over the early part of the fourth century, since the failure of Cephisodotus at Alopeconnesus in 360 was

---

[17] Philoch. Fg. 124. Cf. also the fragment *IG* 2.1².216: τὰ μὲν πομπεῖα . . . Ἀνδροτίων. On the basis of this evidence K. von Fritz, *TAPhA* 71 (1940) 93 is inclined to believe that Androtion was an Exegetes.

[18] There are references in the fragments to each book from I to VIII. Harpocration refers once to Book XII (Fg. 27), but it is probable that the number is incorrect: Ennea Hodoi as an old name for Amphipolis would be more appropriately mentioned in Book II (see p. 80 below) and the mistake ἐν ιβ′ instead of ἐν τῇ β′ is easy (cf. H. Bloch, *HSPh*, Suppl. 1.344f.).

[19] Fg. 14—St. Byz. *s.v.* Ἀργεννοῦσα, νῆσος πρὸς τῇ ἠπείρῳ τῆς Τρωάδος . . . τὸ ἐθνικὸν Ἀργεννούσιος· Ἀνδροτίων ἐν τετάρτῳ τῆς Ἀτθίδος διὰ τοῦ ῑ.

recounted in Book V (Fg. 17). The later books, therefore, even if there are no more than eight books altogether, were evidently much more detailed. Didymus quotes the seventh book for an incident of 350–49; [20] but there is no way of telling how the subject matter was distributed in this part of his work. The account of the origin of the *Bouphonia*, "a very ancient festival of the Athenians," which is attributed to Book IV (Fg. 13), evidently occurred in a digression, a type of digression which recalls the manner of Thucydides.

Another link with Thucydides also suggests itself. Thucydides found occasion to mention the part he himself had played in failing to relieve Amphipolis [21] and his knowledge of some regions at first hand is also occasionally revealed in his writing. It seems that Androtion likewise made some allusions to his own career and did not hide his political sympathies. It has already been suggested that his mention of the cities of Amorgos may be connected with the decree voted in his honour by the people of Arcesine and that he found an opportunity to defend himself against the charge of misappropriating the sacred processional utensils belonging to the state.[22] Another fragment from Harpocration mentions a certain Molpis as one of the ten men who held authority in Peiraeus after the fall of the Thirty.[23] This Molpis is not otherwise known, except that Lysias mentioned him in a lost speech; [24] but since Androtion's own father, Andron, was one of the Four Hundred,[25] it is not surprising that he should have special knowledge of the political events of the end of the fifth century and have friends prominent in oligarchic circles. Since he is cited as an authority for the developments in Athens after the fall of the Thirty and the appointment of the board of ten,[26] it seems quite possible that he gave prominence in his description to the activities of his own family. Oligarchic associations would also explain his interest in Thucydides, the son of Melesias; Theopompus had called this opponent of

[20] Col. 14, 35–49. This passage, which gives the accounts of Androtion and Philochorus of the question of the Megarian ἱερὰ ὀργάς and its settlement by the Athenians in 350–49 B.C., will be discussed more fully in Chapter 6, pp. 128–29 below.

[21] 4.104–06.

[22] See above pp. 78–79. For a possible allusion to one of his own speeches see note 15 above.

[23] Harp. s.v. Μόλπις· . . . Μόλπις ὁ τῶν ἐν Πειραιεῖ. οἱ δ᾽ ἄρα μετὰ τοὺς τριάκοντα δέκα ἄρχοντες ἦρχον ἐν Πειραιεῖ· ὧν εἷς ἦν ὁ Μόλπις, ὡς Ἀνδροτίων ἐν τρίτῃ Ἀτθίδος.

[24] Lys. Fg. 31 (Thalheim).

[25] Harp. s.v. Ἄνδρων.

[26] Fg. 10—Harp. s.v. δέκα καὶ δεκαδοῦχος.

Pericles son of Pantaenus, and Androtion, not content with merely pointing out his mistake, cleared up all difficulty by distinguishing four different men of that name, including a poet, the son of Ariston.[27]

Though the political side of Athenian history is far better represented in the fragments of Androtion than in those of any other Atthidographer, it is clear that religious matters were not entirely neglected. He mentioned an ancient custom of the Athenians not to sacrifice a sheep if it had not been shorn or had not borne a lamb.[28] He explained that Dionysus obtained his name of *Brisaeus* from Brisa in Lesbos where he had a temple; and he explained the origin of the Athenian festival called *Bouphonia*—how an ox gobbled up the sacrificial cake at the *Diipolia* and was accordingly killed on the spot by a certain Thaulon.[29] He attributed the foundation of the Panathenaic festival to Erichthonius, just as Hellanicus had done.[30] Particularly interesting are his remarks about Eumolpus: "Androtion says," writes a scholiast on Sophocles, "that it was not the first Eumolpus who started the practice of initiation into the mysteries, but another Eumolpus, four generations later than this one; that Eumolpus had a son Ceryx, whose son was called Eumolpus; his son again was Musaeus the poet, and Musaeus' son was the Eumolpus who started the mystic rites and became Hierophant." [31] This duplication of a legendary character, for one purpose or another, was a favourite device of Hellanicus and it certainly looks as though Androtion followed in his footsteps. A third example of his loyalty to the author of the first *Atthis* is perhaps to be found in his mention of Parparon, a little town in Aeolis whose principal claim to distinction was that Hellanicus died there.[32] Since he

---

[27] Fg. 43—Sch. Ar. V. 941: Θουκυδίδης Μελησίου υἱὸς Περικλεῖ ἀντιπολιτευόμενος . . . Θεόπομπος μέντοι ὁ ἱστορικὸς τὸν Παντάινου φησὶν ἀντιπολιτεύεσθαι Περικλεῖ, ἀλλ' οὐκ 'Ανδροτίων, ἀλλὰ καὶ αὐτὸς τὸν Μελησίου. For the different men called Thucydides see Fg. 44—Marcellin. *Vit. Thuc.* Androtion also differed with Theopompus over the name of Hyperbolus' father (Fg. 48—Theopomp. F.95a, in *FGrH* 2 B), and an ostrakon from the Athenian agora has shown that Androtion was right (Bloch, *loc. cit.* 354).

[28] Fg. 41.

[29] Fg. 59, 13.

[30] Fg. 1—Harp. *s.v.* Παναθήναια· . . . ἤγαγε δὲ τὴν ἑορτὴν ὁ 'Εριχθόνιος ὁ 'Ηφαίστου, καθά φασιν 'Ελλάνικός τε καὶ 'Ανδροτίων, ἑκάτερος ἐν πρώτῃ 'Ατθίδος.

[31] Fg. 34—Sch. Soph. *OC* 1053 (accepting the emendation 'Ανδροτίων for "Ανδρων). Hellanicus wrote about the Hierophants in his second book (F.45).

[32] Fg. 9—St. Byz. *s.v.* Παρπάρων, χωρίον ἐν 'Ασίᾳ Αἰολικόν . . . ὁ πολίτης Παρπαρώνιος. . . . 'Ανδροτίων δ' ἐν τρίτῳ 'Ατθίδος Παρπαρωνιώτας φησίν.

mentioned the place in Book III, which would cover the lifetime of Hellanicus, it is quite possible that he recorded the time and place of his predecessor's death.[33]

Mythology also received some attention from him, although his continuous account of mythical times did not go beyond Book I. There is one excellent example of his rationalism: he entirely rejected the miraculous stories of the *Spartoi* at Thebes, and insisted that this name arose because Cadmus and his companions were "scattered wanderers" (σποράδες).[34]   In similar style he maintained that the Amphictyons were originally called ἀμφικτίονες ("dwellers around"), and that the story of Amphictyon, son of Deucalion, was without foundation; and because of the traditional illiteracy of the Thracians, he denied that Orpheus could have been both a sage and a Thracian.[35]   On the other hand, he seems to have told the story of Oedipus in an orthodox manner and to have accepted the pious etymology of Colonus ἱππεύς: that Poseidon first harnessed horses at that place.[36]   Indeed, there is no example in his fragments of rationalist methods applied to Athenian mythology.

There are, however, some signs that he gave unorthodox and rationalistic explanations of the origin of political institutions.   His statement that *Apodektai* were substituted for *Kolakretai* by Cleisthenes seems to be simply a mistake, since inscriptions attest the activity of *Kolakretai* well on into the fifth century.[37]   But he also remarked that the *Kolakretai* (in the time of Solon presumably) provided the envoys going to Delphi with money for their journey ("or for any other purpose that might be necessary") out of τὰ ναυκληρικά.[38]   Aristotle points out that in the old laws of Solon there frequently occur phrases like τοὺς ναυκράρους εἰσπράττειν and

---

[33] Perhaps the Halicarnassian whom he mentioned in the same book was Herodotus. Cf. Fg. 6—St. Byz. *s.v.* ῾Αλικάρνασσος· . . . ὁ πολίτης ῾Αλικαρνασσεύς. . . . ᾿Ανδροτίων δ᾽ ἐν τρίτῃ ᾿Ατθίδος ῾Αλικαρνάσσιός φησι.

[34] Fg. 28—30.

[35] Fg. 33—36.

[36] Fg. 31, 32.

[37] Fg. 3—Harp. *s.v.* ἀποδέκται· ἀρχή τίς ἐστι παρ᾿ ᾿Αθηναίοις οἱ ἀποδέκται . . . ὅτι δὲ ἀντὶ τῶν κωλακρετῶν οἱ ἀποδέκται ὑπὸ Κλεισθένους ἀπεδείχθησαν ᾿Ανδροτίων β΄. For the evidence of the inscriptions see J. Oehler, *RE s.v.* Κωλακρέται.

[38] Fg. 4—Sch. Ar. *Aν.* 1540: τὸν κωλακρέτην, τὸν ταμίαν τῶν πολιτικῶν χρημάτων. ᾿Αριστοφάνης ὁ γραμματικὸς τούτους ταμίας εἶναί φησι τοῦ δικαστικοῦ μισθοῦ, οὐ μόνον δὲ τούτου τὴν ἐπιμέλειαν ἐποιοῦντο, ὥς φησι, ἀλλὰ καὶ τὰ ἐς θεοὺς ἀναλισκόμενα διὰ τούτων ἀνηλίσκετο, ὡς ᾿Ανδροτίων γράφει οὕτως· "Τοῖς δὲ ἰοῦσι Πυθῶδε θεωροῖς τοὺς κωλακρέτας διδόναι ἐκ τῶν ναυκληρικῶν ἐφόδιον ἀργύρια καὶ εἰς ἄλλο ὅτι ἂν δέῃ ἀναλῶσαι."

ἀναλίσκειν ἐκ τοῦ ναυκραρικοῦ ἀργυρίου.[39] Evidently τὰ ναυκληρικά is a "modern spelling" for τὰ ναυκραρικά, and Androtion thought that ναύκραρος was an old form of ναύκληρος. His statement about the *Kolakretai*, therefore, must follow upon a discussion of the Naucraries, whose activities he connected, like Cleidemus, with ships and shipping.[40] Unluckily there is insufficient evidence to reconstruct the whole of his argument and that of Cleidemus on the subject.

More distinctly unorthodox is his account of the Seisachtheia. According to Plutarch, Androtion was one of the authors who denied that Solon cancelled all debts by this measure; his view was that interest rates were reduced and that the "shaking off of the burden" consisted in this concession, together with alterations in the weights and measures which had the effect of an inflation.[41] It seems fairly clear that Aristotle is deliberately rejecting this view when he remarks that "Solon's cancellation of debts preceded his legislation and his changes in weights and measures were subsequent." [42]

In other matters, however, there is good evidence that Aristotle found Androtion a useful source of information. Androtion's "rationalization" of the Seisachtheia is in conformity with his position as a "moderate" in politics, who looked back to Solon for his political ideals; it would be natural for him to make this measure appear less revolutionary and more constitutional than the traditional view represented. It is quite probable, therefore, that the

[39] *Resp. Ath.* 8.3.

[40] See section on Cleidemus, pp. 58–59, 67–68 above.

[41] Fg. 40—Plu. *Sol.* 15: καίτοι τινὲς ἔγραψαν, ὧν ἐστιν ᾿Ανδροτίων, οὐκ ἀποκοπῇ χρεῶν, ἀλλὰ τόκων μετριότητι κουφισθέντας ἀγαπῆσαι τοὺς πένητας, καὶ σεισάχθειαν ὀνομάσαι τὸ φιλανθρώπευμα τοῦτο, καὶ τὴν ἅμα τούτῳ γενομένην τῶν τε μέτρων ἐπαύξησιν καὶ τοῦ νομίσματος ἐς τιμήν.

[42] *Resp. Ath.* 10: ἐν μὲν οὖν τοῖς νόμοις ταῦτα δοκεῖ θεῖναι δημοτικά, πρὸ δὲ τῆς νομοθεσίας ποιῆσαι τὴν τῶν χρεῶν ἀποκοπὴν καὶ μετὰ ταῦτα τήν τε τῶν μέτρων καὶ σταθμῶν καὶ τὴν τοῦ νομίσματος αὔξησιν. Cf. B. Keil, *Die Solonische Verfassung* 45–46. N. G. L. Hammond, "The Seisachtheia and the Nomothesia of Solon," *JHS* 60 (1940) 78, calls this passage a "tacit criticism of Androtion's theory"; but on p. 75 he writes: "As *Atthides* were written in a chronological form, it is clear what Aristotle has done with the work of Androtion; he has re-arranged the matter in a form suitable to his purpose, laying emphasis on constitutional points and passing his verdicts on Solon the constitutionalist, and he has then introduced a note on chronology based on Androtion's *Atthis*." This is an impossible conclusion, since Androtion denied the χρεῶν ἀποκοπή altogether. Hammond is also quite unjustified in his assumptions that a chronological (does he mean annalistic?) form was customary in *all* portions of *Atthides* and that Androtion's account "crystallized" fourth century tradition.

idea of Solon as ὁ μέσος πολίτης, which is stressed by Aristotle, derives not only from the Athenian "moderates" in general, but from Androtion in particular.[43] One notices a similar leaning to the "moderates" in Aristotle's later chapters, especially in his summary of the various προστάται at Athens in chapter 28, where Nicias, Thucydides, and Theramenes are said to be "the best of the Athenian politicians after those of the early days." There are, moreover, a number of points of agreement with Androtion over matters of fact: Hipparchus was the first victim of ostracism, and Peisistratus on his second return from exile won his victory in the deme Palleneis.[44] Again, Androtion is said to have described "what followed" after the appointment of a committee of ten subsequent to the fall of the Thirty; and Aristotle describes these events in detail.[45] When this evidence is taken into account, the case for identifying Androtion the historian with the orator seems almost complete.[46]

The fragments of the earlier Atthidographers revealed no political bias and indeed no very lively interest in political questions. Although this may be an accident, the lack of any evidence makes it impossible to argue that there was a consistent political tradition which they followed. There is no reason to suppose that, because Androtion was a "moderate," his predecessors held similar political views, and consequently there are really no grounds for believing in the existence of an "Atthid tradition" of political history; indeed, as we have seen, there were questions of fact on which the different writers were not agreed.

There is, unfortunately, no evidence to show whether Androtion revealed any political opinions in his account of the history of the fourth century. In his account of the fifth century, however, he took occasion to note the exact year in which Cleon died and remarked on the ostracism of Hyperbolus "as an undesirable" (διὰ φαυλότητα).[47] He also described how the Athenian general, Phormio, after an honest term of office as strategus, was in poverty and suffered *atimia* through inability to pay a debt to the treasury; and

[43] On this point cf. F. E. Adcock, *Kl* 12 (1912) 14–16.

[44] 22.4—Fg. 5; 15.3—Fg. 42.

[45] 38—Fg. 10.

[46] De Sanctis in his earlier essay (see bibliography) denied the identification, though admitting A. to be a moderate and perhaps a younger member of the orator's family. In his later essay, after the publication of the Didymus papyrus, he gave way.

[47] Fg. 46, 48.

how the people paid his debt for him, so that he could accept the invitation of the Acarnanians to conduct an Athenian expedition into their country.[48] This tale would be of no particular importance, were it not for the fact that Pausanias (who records the story without understanding all the details) describes Phormio as a member of a distinguished family, "resembling the better class of people in Athens." [49]

Pausanias also tells us that Androtion described the condemnation of the athlete Dorieus by the Spartans, and he remarks that this description seems like an attempt to compare the action of the Spartans with the Athenians' treatment of their generals after Arginusae.[50] This may possibly be an indication of anti-Spartan animosity; but we also learn that he did not attempt to disguise the defeat suffered by the Athenians in the territory of Corinth and that he described the condemnation of Cephisodotus for his failure at Alopeconnesus.[51] These fragments, however, give no indication of the general character of this part of his work; they do not even show whether he used an annalistic system for events of his own lifetime; the only date which is referred to an archon's name is that of Cleon's death, in the archonship of Alcaeus, two years after the production of the *Clouds*.

With so little evidence available about Androtion's treatment of fourth century affairs, there is not much to support the view of De Sanctis that he is the author of the *Hellenica* of Oxyrhynchus. In fact, apart from the mere possibility on chronological grounds, the only real argument of De Sanctis is that the author of the *Hellenica* is "what we should call a moderate, and furthermore a man of action, one who would examine the reasons of expediency which seem to him to determine men's actions and would devote himself extensively to financial questions." [52] He also insists on the interest shown by Androtion in extra-Athenian affairs—a tend-

---

[48] Sch. Ar. *Pax* 347 (not in Müller *FHG*).

[49] Paus. 1.23.10: Φορμίωνι γὰρ τοῖς ἐπιεικέσιν ᾿Αθηναίων ὄντι ὁμοίῳ καὶ ἐς προγόνων δόξαν οὐκ ἀφανεῖ συνέβαινεν, κτλ. It is only the scholiast on Aristophanes, not Pausanias, who gives Androtion as his authority. Bloch, *loc. cit.* 348–51, sets the two passages side by side and takes the errors of Pausanias as evidence that he did not use Androtion's *Atthis* directly; but he says nothing about the political significance of the story of Phormio.

[50] Fg. 49—Paus. 6.7.6–7.

[51] Fg. 50, 17.

[52] *AAT* 43 (1907–08) 348. Bloch (*loc. cit.* 328–34) gives other reasons for rejecting the view of De Sanctis.

ency which is suggested only by the geographical fragments; no fragment reveals an interest in political events in which Athens was not concerned.   Since this is the case and since the points of similarity between the Oxyrhynchus extract and the fragments of the Atthidographers are so few, we must renounce any attempt to draw conclusions about the *Atthides* from that quarter.

BIBLIOGRAPHY

The Fragments

C. G. Lenz and C. G. Siebelis, *Philochori Atheniensis librorum fragmenta; accedunt Androtionis* Ἀτθίδος *reliquiae* (Leipzig, 1811) 109–22.

C. and T. Müller, *Fragmenta Historicorum Graecorum (FHG)* 1.lxxxiii–iv, lxxxviii, 371–77; 4.645–46.

R. Stiehle, "Zu den Fragmenten der griech. Historiker," *Ph* 8 (1853) 634–36.

A. Bauer, "Zu den Fragmenta Historicorum Graecorum," *WS* 5 (1883) 157.

H. Usener, "Ein Fragment des Androtion," *Jahrb. class. Phil.* 103 (1871) 311–16.

Didymus, *Commentary on speeches of Demosthenes*, first published by H. Diels and W. Schubart, *Berliner Classikertexte* 1 (1904) 1–95; afterwards in Teubner edition, *Volumen Aegypticum* 4.1.

Discussion

E. Schwartz, *RE s.v.* "Androtion."

F. E. Adcock, "The source of the Solonian chapters of the Athenaion Politeia," *Kl* 12 (1912) 1–16.

H. Bloch, "Studies in historical literature of the fourth century B.C.," *HSPh*, Suppl. 1, Special Volume in Honour of W. S. Ferguson (1941) 303–55.

I. Bywater and V. Rose, "Anonymus zu Aristot. Ethica V," *H* 5 (1871) 354–59.

G. De Sanctis, "Studi sull' Ἀθ. Πολ. attribuita ad Aristotele," *RFIC* 20 (1899–92) 147–63.

Id., "L'Attide di Androzione e un papiro di Oxyrhynchos," *AAT* 43 (1907–08) 331–56.

M. Heller, *Quibus auctoribus Aristoteles in Rep. Ath. conscribenda et qua ratione usus sit* (Diss. Berlin, 1893).

B. Keil, *Die Solonische Verfassung in Aristoteles' Verfassungsgeschichte Athens* (Berlin, 1892).

A. Momigliano, "Androzione e le 'Elleniche' di Ossirinco," *AAT* 66 (1931) 29–49.

Th. Reinach, "Zu Andariton Fr. 40," *H* 63 (1928) 238–40.

Id., "Une correction d'un texte de l'annaliste Androtion cité par Plutarque," *CRAI* 1927, 318.   (Summary of the article cited above.)

F. W. Schneidewin, "Zu Apsines und Androtion," *Ph* 5 (1850) 237.

IV. SOME LESSER FIGURES

Before going on to deal with Philochorus and Ister, about whom far more is known, it will be convenient to set forth briefly the scanty information that is available about three shadowy Athenian figures who are generally regarded as Atthidographers: Melesagoras, Demon, and Melanthius.

But first a word must be said about Andron of Halicarnassus, to whom Müller attributes an *Atthis*.[1] No ancient writer ever refers to his work under this title; but he wrote a work called Συγγένειαι (*Families*), a title which recalls the *Genealogies* of Hecataeus, and some of the fragments show that he touched on various points of Attic legend. Strabo remarks that, although the Atthidographers disagree on many questions, "all that are of any account" agree that in the division of land between the four sons of Pandion Nisus obtained the Megarid; and then he adds in parenthesis the divergent views of Philochorus and Andron on the extent of Nisus' portion.[2] There are other fragments of Andron referring to Eumolpus and the founding of the Isthmian games by Theseus;[3] and Harpocration, in his note about the Phorbanteum, quotes Hellanicus and "Andron in the eighth book of his Συγγένειαι."[4] Müller thought that these fragments must come from an *Atthis*, which corresponded to the eighth book of a comprehensive mythographical and historical work.

The only evidence that Andron carried his treatment of Attic affairs down to historical times is a remark of the scholiast on the *Frogs* that "Andron differed from Xenophon about the recall of Alcibiades."[5] Jacoby, however, rejects even this indication and thinks that the scholiast intended to cite Androtion instead of Andron. In any case, with so little evidence available, further discussion is not profitable. Since Andron's work is never cited as an *Atthis*, the mere fact that he discussed Attic myths in a general mythographical work does not entitle him to be regarded as an Atthidographer. His date is quite uncertain; Jacoby thinks that he may belong to the fourth century.

Melesagoras[6] is another problematic figure. Even his name is uncertain, since in the manuscripts of some authorities it appears as Amelesagoras. Schwartz[7] wanted to explain this very curious name as derived from the river of the underworld called Ameles

---

[1] For the fragments see Müller, *FHG* 2.349-52, Jacoby, *FGrH* 1. no. 10. References will be given to the latter collection. See also E. Schwartz, *RE s.v.* "Andron" (11).

[2] F. 14—Str. 9.1.6.

[3] F. 6, 13—Plu. *Thes.* 25, Sch. Soph. *OC* 1053 (where the emendation Ἀνδροτίων for Ἄνδρων is probably correct). See above p. 81, and note 31.

[4] F. 1—Harp. and Suid. *s.v.* Φορβαντεῖον.

[5] F. 18—Sch. Ar. *Ra.* 1422. See Jacoby's note and Müller *FHG* 2.346.

[6] For the fragments see *FHG* 2.21-22.

[7] *RE s.v.* "Amelesagoras." Another Amelesagoras is mentioned in some MSS. in [Hippoc.] *Ep.* 11 (ed. Littré 9.324).

(River of Freedom from Care), but Kroll [8] insists, probably rightly, that this is an impossible name and must be, wherever it occurs, a mistake for Melesagoras.

A more serious problem is that of his date. Dionysius of Halicarnassus, in his well known list of "early historians before the Peloponnesian War," groups him with Acusilaus of Argos and Charon of Lampsacus; Clement of Alexandria, when he is discussing plagiarism among the Greek historians, says that Gorgias of Leontini and Eudemus of Naxos and a number of logographers and Atthidographers stole material from him; while Maximus of Tyre speaks of him as an Eleusinian "inspired prophet" (κάτοχος θείᾳ μοίρᾳ) and clearly thinks that he belongs to quite early times.[9] If we accept the statement of Dionysius of Halicarnassus, we must suppose that Melesagoras was the first author to write an *Atthis*.[10] Furthermore, we must believe that he preceded Hellanicus in working out the longer list of kings between Cecrops and Theseus and was the first to invent Erichthonius as distinct from Erechtheus, since the story of the birth of Erichthonius is told in one of his fragments. This fragment [11] tells the tale (afterwards told by Callimachus in the *Hecale*) [12] of Athena's deadly quarrel with crows: the infant Erichthonius, the "earth-born" child of Hephaestus, had been hidden inside a box and Athena had entrusted this to the daughters of Cecrops with strict orders that they were not to open it. Meantime she left Athens; and when she was on her way back, carrying a mountain which was to be set in front of the Acropolis so as to increase the natural defences of Athens, a crow met her with the news that the daughters of Cecrops had disobeyed her order; whereupon, in her anger at receiving this bad news, she forbade crows to perch on the Acropolis henceforward and dropped the mountain, which was subsequently named Mount Lycabettus.

It seems fairly clear that Callimachus took this story from Melesagoras, just as he borrowed other tales from various Atthi-

[8] *RE s.v.* "Melesagoras." This Greek author thus enjoys the unusual honour of two articles in *RE*.

[9] D.H. *Th.* 5; Clem. Al. *Strom.* 6.2.26 (ed. Stählin 443); Max. Tyr. 38.3 (ed. Hobein 439).

[10] For a statement of this view see M. Wellmann, "Beitrag zur Geschichte der attischen Königsliste," *H* 45 (1910) 554–63.

[11] Fg. 1—Antig. *Mir.* 12. The text runs: Ἀμελησαγόρας δὲ ὁ Ἀθηναῖος ὁ τὴν Ἀτθίδα συγγεγραφώς. This is the only citation of the title of the work.

[12] Ida Kapp, *Callimachi Hecalae Fragmenta*, Fg. 60–62; R. Pfeiffer, *Callimachi fragmenta*, Fg. 34, pp. 80–81. Cf. Wilamowitz, *Phil. Untersuch.* 4.24.

dographers.   But his use of the story does not prove that Melesagoras belongs to the early part of the fifth century rather than to
the latter part of the fourth.[13]   The early date given by Dionysius
should arouse suspicion because no local history of Attica is heard
of at that time; that particular literary form does not occur in
the period before the Peloponnesian War.[14]   On the other hand,
Clement's remarks on plagiarism need not be taken seriously (his
statements on this subject contain more than one chronological
impossibility), and Maximus of Tyre is evidently thinking of some
legendary or semi-legendary personage.   No elaborate theories
about forgery are necessary to explain their statements; and if we
suppose that Dionysius has been misled into dating Melesagoras
half a century too early (perhaps by the same authorities who were
responsible for the misdating of Hellanicus), there is no difficulty
in accepting the fragments as belonging to a genuine work written
in the fifth or fourth century.

There are, indeed, two fragments which are most appropriate
to an *Atthis*, since they relate to the Cretan adventures of Theseus.
Hesychius says that he gave Androgeos the name of Eurygyes;
and, according to his story, the blinding of Asclepius was due to his
resuscitation of Glaucus, the son of Minos.[15]

Little more than this can be said about the *Atthis* of Demon.[16]
Plutarch cites both Demon and Philochorus for the view that the
Minotaur was not a monster, but simply a general of Minos called
Taurus, though their versions of Theseus' triumph are somewhat
different.[17]   Suidas says that Philochorus wrote his *Atthis* "in reply
to Demon," and a special work of his *Against the Atthis of Demon*
is also attested.[18]   The natural interpretation of this evidence is
that Demon was an elder contemporary of Philochorus and anticipated him in his rationalization of the Minotaur legend.   Of the
three other fragments from the *Atthis*, one refers to the Tritopatores
as winds (recalling the Orphic doctrine that they were guardians

---

[13] Wellmann (*loc. cit.* 560) thinks Callimachus used a work written in the fifth
century and that there was a later forgery (made in early imperial times) attributed to
an "Eleusinian seer."   Wilamowitz prefers the fourth century and is followed by
W. Schmid, *Griech. Literaturgesch.* 1.1.707f.   Cf. also Müller, *FHG* 2.22 and Susemihl,
Gesch der griech. Lit. in der Alexandrinerzeit 1.599.

[14] Cf. Leo Weber, "Nachträgliches zu Androgeos," *RhM* 78 (1929) 26–29.

[15] Fg. 3—Hsch. *s.v.* ἐπ' Εὐρυγύῃ ἀγών; Fg. 2—Apollod. 3.10.3; Sch. Eur. *Alc.* 1.

[16] For the fragments, see *FHG* 1.378–83.   Cf. also E. Schwartz, *RE s.v.* "Demon"
(6).

[17] Fg. 3—Plu. *Thes.* 19.   See below, Chap. 7, p. 152.

[18] See below, chap. 6, p. 108.

of the winds), and another refers to the institution of the Oschophoria by Theseus and tells how the procession at that festival was
supposed to be a reënactment of the return of the Athenian boys
and girls from Crete.[19]   Both these fragments recall Demon's supposed controversy with Philochorus, who maintained that the Tritopatores were children of Ge and Uranus (and hence the "third
fathers" of the human race) and also had something to say about
the Oschophoria.[20]

The remaining fragment is quoted by Athenaeus from the fourth
book of the *Atthis*:[21] "He says that when Apheidas' was king of
Athens, Thymoetes, his younger brother, who was a bastard son,
killed him and himself became king.   In his reign Melanthus, a
Messenian, being exiled from his country, inquired of the Pythian
where he should settle; she said that he should stay in that place
where he was received with hospitality and his hosts served him
at dinner with the feet and the head of an animal.   And this
happened to him at Eleusis; since the priestesses were celebrating
a local festival and had used up all the meat and only the feet and
head were left, they offered these to Melanthus."   Melanthus is
father of Codrus, the famous last king of Athens, so that this
fragment gives Demon's sequence for the last four Athenian kings.
Nothing is known of his treatment of later times.

More numerous than the fragments from Demon's *Atthis* are
those from his work *On Proverbs*,[22] which is cited by scholiasts and
lexicographers to explain the origin of various proverbial sayings.
There is also one fragment from a work *On Sacrifices*.[23]   Evidently
Demon conformed to type and, like the other Atthidographers, was
interested in religious antiquities.

Finally something must be said about Melanthius.[24]   There is
only one reference to his *Atthis*, which is not particularly informative,[25] but there are also three fragments from his work *On the
Mysteries at Eleusis:*[26] one of these describes the details of sacrifice

[19] Fg. 2—Suid. *s.v.* Τριτοπάτορες; Fg. 4—Plu. *Thes.* 23.

[20] Fg. 2, 3, 44.

[21] Fg. 1—Ath. 3.96D–E.

[22] For which see O. Crusius, *Analecta critica ad paroemiographos Graecos* (Leipzig,
1883) 132–50; *Ph*, Suppl. 6 (1891–93) 269–74.

[23] Fg. 22—Harp. *s.v.* προκώνια.

[24] For the fragments, see *FHG* 4.444.

[25] Fg. 1—Harp. *s.v.* γρυπάνιον.   He is cited as using the phrase καὶ ἔγρυπεν ἡ γῆ
in describing an earthquake.

[26] Fg. 2–5 (4 and 5 really constitute one fg.).

to Hecate, another the customs which governed initiation into the mysteries, and the third the case of Diagoras of Melos, mentioned in the *Birds* of Aristophanes: "And if one of you shall kill Diagoras the Melian he shall have a talent." [27] The scholia on this passage record that, according to Melanthius, this man disparaged the mysteries and indeed deterred many from being initiated; so that a stele was put up denouncing him and the citizens of Pellene who refused to give him up, in which the following clauses were included: ἐὰν δέ τις ἀποκτείνῃ Διαγόραν τὸν Μήλιον, λαμβάνειν ἀργυρίου τάλαντον· ἐὰν δέ τις ζῶντα ἀγάγῃ, λαμβάνειν δύο. The scholiast quotes the actual words of the decree from his work and, in referring to Craterus, gives only a paraphrase of it; hence Wilamowitz decided that Melanthius quoted the actual words and Craterus did not; and that therefore Melanthius must be earlier than Craterus—a very curious piece of misleading argumentation.[28] A number of respected Athenian citizens called Melanthius, some of them holders of priestly offices,[29] are mentioned in inscriptions from the fifth and fourth centuries, but without external evidence it is idle to attempt an identification.

Apart from the fragments given by Müller it is possible that one further reference to the work on the mysteries is to be found.[30] But in any case the evidence available is very slight. The fragments, such as they are, seem to show that he conformed to type and they do not suggest that his work on the mysteries was other than a dignified work, such as might be written by a priestly official, showing proper respect for the secrets that might not be revealed. His quotation of the actual words of a decree suggests a comparison with the work of Philochorus on *Attic Inscriptions*.

Such information, therefore, as is available about Melesagoras, Demon, and Melanthius adds little to our knowledge of the Atthis tradition. It does show, however, that, besides those writers of whose works more numerous fragments survive, there may have been others of similar tastes writing works of a similar kind. It does not tell us in what respects their work may have differed from that of their more illustrious colleagues.

[27] 1073f.

[28] *Aristoteles und Athen* 1.286f.

[29] Cf. Kirchner, *Prosop. Attica s.v.* "Melanthios."

[30] Cf. Andrée, *RE s.v.* "Melanthios" (11), who refers to Sch. Ap. Rhod. 1.1126 (where the MSS. have Μαιάνδρῳ or Μενάνδρῳ).

CHAPTER V

EPHORUS, THEOPOMPUS, AND ARISTOTLE

Since many of the characteristics of Hellanicus were found to be present in the work of his contemporary Thucydides, it seems appropriate at this point to attempt some comparison between the *Atthides* written in the fourth century and other historical work produced at the same time. A comparison of this kind is bound to be less conclusive than that which was attempted in Chapter 2, because the works of the most distinguished historians of the later fourth century have been lost; instead of comparing fragments with a complete extant work, we are obliged in this case to compare fragments with fragments. Even Aristotle's *Constitution of Athens* is not a complete extant work in the same sense as the history of Thucydides. Although it is preserved in almost complete form, it is in many ways comparable to a collection of fragments; it is the only surviving example of a series of treatises on constitutional history; it is not a self-contained work of Aristotle, but a fragment of the work of his pupils; and its special qualities could be better understood if we possessed the entire Encyclopaedia of Constitutions compiled by the master and his students. At the same time, Ephorus and Theopompus are sufficiently well represented in their fragments, so that we are able not only to find points of similarity between their books and the *Atthides*, but also to recognize some of their distinctive characteristics. But the special peculiarities of the pupils of Isocrates [1] and Aristotle cannot occupy our attention here except in passing; this chapter does not claim to offer a full critical treatment of their work, but only to emphasize what they have in common with the Atthidographers.

When the fragments of Ephorus are taken together with those portions of Diodorus which depend on his work, there is an abundance of material relevant to the present discussion.[2] Certain tradi-

---

[1] There seems to be no adequate reason to reject the tradition that Ephorus and Theopompus were pupils of Isocrates (despite the arguments of E. Schwartz, *RE s.v.* "Ephoros," 6.1–16). For discussion, see A. E. Kalischek, *De Ephoro et Theopompo Isocratis discipulis* (Diss. Münster, 1913).

[2] The fragments will be quoted from Jacoby's collection, *FGrH* 2A, no. 70. Jacoby makes no attempt to include all passages of Diodorus and other authors who may be borrowing from Ephorus. Cf. his remarks in *FGrH* 3A, Intro. 8*.

tional features, inherited from old Ionian historiography, are illustrated remarkably well. There is plentiful evidence of his interest in geography, which he seems to have indulged almost as freely as Herodotus.[3] Equally evident is his love of digressions, when an opportunity occurs to write about "foundings of cities and their founders, migrations, family histories";[4] numerous citations by Stephanus of Byzantium and others bear witness to his interest in *Ktiseis*.[5] His pride in his native city of Cyme and his anxiety to remind his readers of the part it played in Greek history has laid him open to much unkind criticism.[6] But extravagant local patriotism is not confined to Ephorus; we have found examples of it in the *Atthides*, and there is no reason why it should be confined to Athenian writers.

His interest in myths and genealogy is equally evident. Although he professed to begin his history with the return of the Heraclidae,[7] this formal limitation of his theme did not deter him from all kinds of digressions. We find references to the story of Heracles, to the settlement at Delphi with criticism of the legends about Apollo, to the legendary traditions of Aetolia, to the story of Minos, which he rationalized.[8] His interest in Homeric questions was not confined to his *Local History* ('Επιχώριος Λόγος) of his native Cyme, but he discussed the Homeric Ethiopians and Cimmerians in his geographical books (IV–V), and placed the latter in the caves near Cumae rather than in their traditional northern habitat.[9] On the other hand, there is a noteworthy lack of allusions among

[3] Cf. J. Forderer, *Ephoros und Strabon* (Diss. Tübingen, 1913), and M. Rostovtzeff, *Skythien und der Bosporus* 6–7, 80–86.

[4] T. 18a—Str. 10.3.5: Πολύβιος. . . . φήσας περὶ τῶν Ἑλληνικῶν καλῶς μὲν Εὔδοξον, κάλλιστα δ' Ἔφορον ἐξηγεῖσθαι περὶ κτίσεων, συγγενειῶν, μεταναστάσεων, ἀρχηγετῶν. T. 18b—Plb. 9.1.4: τὸν μὲν γὰρ φιλήκοον ὁ γενεαλογικὸς τρόπος ἐπισπᾶται, τὸν δὲ πολυπράγμονα καὶ περιττὸν ὁ περὶ τὰς ἀποικίας καὶ κτίσεις καὶ συγγενείας, καθά που καὶ παρ' Ἐφόρῳ λέγεται, τὸν δὲ πολιτικὸν ὁ περὶ τὰς πράξεις τῶν ἐθνῶν καὶ πόλεων καὶ δυναστῶν.

[5] Cf. F. 11, 18b, 21, 24, 31b, 39, 40, 44, 56, 78, 89, 115, 126, 127, 136, 137, 146, 164, 216.

[6] Schwartz (*loc. cit.*) is particularly severe in speaking of his "small town" outlook on history.

[7] T. 8—D.S. 4.1.3: Ἔφορος μὲν γὰρ ὁ Κυμαῖος, Ἰσοκράτους ὢν μαθητής, ὑποστησάμενος γράφειν τὰς κοινὰς πράξεις, τὰς μὲν παλαιὰς μυθολογίας ὑπερέβη, τὰ δ' ἀπὸ τῆς Ἡρακλειδῶν καθόδου πραχθέντα συνταξάμενος ταύτην ἀρχὴν ἐποιήσατο τῆς ἱστορίας. Cf. 16.76.5—T.10. It is uncertain how literally we should interpret Diodorus in such matters, since he speaks of Herodotus as "beginning from the time of the Trojan War."

[8] F.13–15, 31b, 122, 147.

[9] F.128, 134.

the fragments to Attic myths. Even Theseus appears only once; [10] and since Minos is represented as a just lawgiver instead of a cruel tyrant,[11] the crowning exploit of Theseus' career, the slaying of the Minotaur, is evidently denied altogether. There is no reference to any stories about the Athenian kings either in his fragments or in the fragments of the lost books of Diodorus. There is, therefore, no evidence to show that he played any part in developing the Attic mythological tradition, with which the Atthidographers were so much concerned.

We should be in a better position to understand his method of dealing with local legends if we knew more about his *Local History* of Cyme. But unfortunately—and not unnaturally, since probably it was read by few people outside Cyme—we know nothing of this work except that it claimed Homer and Hesiod as natives of that city. This single scrap of evidence, recorded by a surprisingly large number of ancient authors,[12] gives us no useful clue to its character, its value as an historical work, its system of chronology, the proportion of space it allotted to earlier and later times, its manner and style, or its relation to earlier local histories.[13] Eduard Schwartz insists that this work was a collection of patriotic anecdotes—an assumption quite unjustified by the evidence.[14] But even though we do know so little about its character, it is useful to remember that Ephorus wrote a local history, as well as a universal one. It seems that he and the Atthidographer, who like him is supposed to have been a pupil of Isocrates, had some interests other than those which Isocrates tried to encourage in his students.[15]

But if Ephorus showed comparatively little interest in myths relating to Athens and seems to have paid more attention to other legends, it cannot be said of him that he neglected Athens otherwise or that his work is lacking in other characteristics typical of an

[10] F.23 (the good relations between Athens and Thessaly are traced to the friendship of Theseus and Peirithous).

[11] F.147.

[12] F.1, 97—103.

[13] G. L. Barber, *The Historian Ephorus* (Cambridge, 1935) 4–5, has a violent outburst against local histories of this period and, after citing these fragments, adds: "No better example than this is needed to prove that annals such as these were not an accurate record of local history, but a chronicle more often than not deliberately forged to promote the fame of one's native city." This sweeping condemnation is not justified at all.

[14] *RE* 6.2: "Solche panegyrischen Zusammenstellungen der vaterstädtischen Traditionen und dessen was zur Tradition gemacht wurde."

[15] Cf. Chap. 4, p. 78 above.

Atthis. His bias in favour of Athens, so clear from the text of Diodorus, has been discussed many times [16] and needs no further illustration here. Apart from Diodorus, whose chief interest is in his account of political history, the fragments properly so-called show how much attention he paid to anecdotes about Athenian politicians and to the details of Athenian religious and social institutions. In speaking of the Apaturia he offered an etymological explanation of the name of this festival.[17] After relating how Themistocles was ostracized (the first to suffer this fate, according to his account), he gave a description of the institution of ostracism, ending with the fine sentiment, as reported by Diodorus, that the Athenians "appear to have established this custom not in order to punish ill-doing, but in order that the spirits of the victims may be humbled through exile." [18] In telling the well-known story of how Cimon paid the fine imposed on his father Miltiades, he added what seems like an individual touch: that Cimon had married a rich woman.[19] He also had a special account of his own about the manner in which Alcibiades met his death: how he revealed to Pharnabazus the plot of Cyrus against Artaxerxes and asked for a safe conduct to the king, so as to make his report personally; and Pharnabazus passed on the warning to the king, but had Alcibiades killed to prevent his getting credit for bringing the information.[20]

Another curious and significant fact revealed by the fragments is the interest of Ephorus in proverbs. From such various sources as the Platonic scholia, Macrobius, and the lexicographers we learn his explanations of at least seven different proverbial sayings. Macrobius quotes in full his explanation of how the term "Achelous" came to be used generally for water (though he does not indicate the occasion of this digression).[21] Stephanus of Byzantium also gives the actual words he used in explaining the term ἀναπαριάζειν: how the Parians, when besieged by Miltiades, had agreed to surrender but went back on their word when they mistook a forest fire on Myconos for a beacon signal from Datis; hence ἀναπαριάζειν came to be used for people who violated an agreement.[22] Of the Atthi-

[16] Cf. e.g. Barber, *op. cit.*, chap. 6, "Bias in Ephorus."
[17] F.22. Cf. Hellanic. F.125 and Ister fg. 3, 4 (*FHG*).
[18] D.S. 11.55.
[19] F.64.
[20] F.70.
[21] F.20.
[22] F.63. Five other sayings are explained in F.12, 19, 27, 58, 59. Cf. also F.37, 149, 175, 183.

dographers Demon actually wrote a separate work *On Proverbs*, and Ister is cited for the explanation of two proverbial sayings.[23] The interest of the historians in origins and *aetia* of this kind is simply another example of the antiquarianism of the fourth and third centuries.

It remains to speak of his chronological method. Diodorus at the beginning of his fifth book praises Ephorus for dividing his material into books according to subject matter; and he announces his intention of adopting a similar procedure himself, giving the name *The Book of the Islands* to the book which he is just starting.[24] In his later books, however, beginning with Book XI, Diodorus uses an annalistic arrangement, but he sometimes makes the mistake of crowding into a single year a series of events which extended over a longer period. A notable example of this kind of confusion is in 11.60–62, where all Cimon's movements, from Eion to the Eurymedon, are crowded into the year 469.[25] His error here is very possibly due to careless reading of Ephorus. A papyrus fragment, which has with good reason been identified as the work of Ephorus,[26] describes these campaigns of Cimon without marking the passage of the years; and if this is indeed the source used by Diodorus, it is easy to see how he might make mistakes in trying to fit this kind of narrative into an annalistic arrangement. It is certain that Ephorus devoted some books entirely to western affairs and there is no suggestion that his early books were arranged according to any chronological scheme at all.[27] The papyrus fragment cited above is the best indication of the method he used in dealing with the fifth century; if the *Hellenica* of Oxyrhynchus could be identified with certainty as his work, it would be good evidence that he followed some kind of annalistic system for the fourth century; but there is no other evidence to support such a conclusion. A detailed comparison of his method with that of the Atthidographers or of Thucydides and Xenophon is therefore not possible.

[23] Fg. 1, 2 (*FHG*).

[24] D.S. 5.1.4—T.11: Ἔφορος δὲ τὰς κοινὰς πράξεις ἀναγράφων οὐ μόνον κατὰ τὴν λέξιν, ἀλλὰ καὶ κατὰ τὴν οἰκονομίαν ἐπιτέτευχε· τῶν γὰρ βίβλων ἑκάστην πεποίηκε περιέχειν κατὰ γένος τὰς πράξεις. διόπερ καὶ ἡμεῖς τοῦτο τὸ γένος τοῦ χειρισμοῦ προκρίναντες κατὰ τὸ δυνατὸν ἀντεχόμεθα ταύτης τῆς προαιρέσεως.

[25] His narrative closes with the remark ταῦτα μὲν οὖν ἐπράχθη κατὰ τοῦτον τὸν ἐνιαυτόν. Cf. also 11.55–59, where the closing period of the life of Themistocles is condensed into a single year.

[26] F.191—*POxy* 13.1610.

[27] For the material covered in the different books see Jacoby, *FGrH* 2 C 27–30.

In view of what has already been said about Ephorus, the position of Theopompus [28] with regard to the *Atthides* may be described more briefly. His fragments are more numerous than those of Ephorus, and they illustrate certain of his interests very clearly and fully. Most of these fragments belong to the *Philippica*. His *Hellenica* is generally supposed to be an earlier work, and there is less evidence of bombast and rhetorical method in it than in the *Philippica*. Since it devoted twelve books to a period of seventeen years (411-394 B.C.),[29] and evidence of long digressions is lacking, it presumably treated events in some detail, as indeed the anonymous *Life of Thucydides* testifies.[30] We might reasonably expect to find more touches reminiscent of an *Atthis* here; but we have much less information about it than about the *Philippica*, and only those who include the *Hellenica* of Oxyrhynchus among its fragments have been able to reach definite conclusions about its character; if we reject this identification, there is no evidence available about its chronological method nor the proportion of space it devoted to purely Athenian affairs. Indeed, since traces of the characteristics which concern us in this chapter appear indiscriminately in fragments of the *Hellenica* and the *Philippica*, it will be convenient, for the present purpose, to make no distinction between the two works nor between different periods in the life of the author. The possible differences between these two works, the relation of Theopompus to Isocrates, and the stages of his development from orator to historian cannot be discussed here.

Unlike Ephorus, Theopompus apparently took little interest in mythological tradition except that which related to *Ktiseis*. To the stories of foundations of cities, as appears from the criticism of Dionysius of Halicarnassus, he devoted some attention, and to migratory movements in general.[31] But when he remarked that he would recount μῦθοι in his histories "better than Herodotus and

---

[28] The fragments are in *FGrH* 2 B, no. 115. For discussion see R. Laqueur, *RE s.v.* "Theopompos" (9), 5A.2176–2223, who follows Ed. Meyer and other German critics in believing Theopompus to be the Oxyrhynchus historian; A Momigliano, "Studi sulla storiographia greca del IV secolo a. C. I. Teopompo," *RFIC*, N.S. 9 (1931) 230–42, 335–53; K. von Fritz, "The Historian Theopompus," *AHR* 46 (1941) 765–87.

[29] T.13, 14—D.S. 13.42.5; 14.84.7.

[30] Chap. 5—F.5: καὶ γὰρ τὸ τεῖχος αὐτῶν καθῃρέθη καὶ ἡ τῶν τριάκοντα τυραννὶς κατέστη καὶ πολλαῖς συμφοραῖς περιέπεσεν ἡ πόλις, ἃς ἠκρίβωσε Θεόπομπος.

[31] T.20—D.H. *Pomp.* 6.4: καὶ γὰρ ἐθνῶν εἴρηκεν οἰκισμοὺς καὶ πόλεων κτίσεις ἐπελήλυθε, βασιλέων τε βίους καὶ τρόπων ἰδιώματα δεδήλωκε, καὶ εἴ τι θαυμαστὸν ἢ παράδοξον ἑκάστη γῆ καὶ θάλασσα φέρει.

Ctesias and Hellanicus and the authors of *Indica*," [32] he evidently meant not so much episodes from mythology as tales of the marvellous (θαύματα) and strange new stories. The few mythological allusions that the fragments reveal are distinctly heterodox: for example, his story that Odysseus, after his reunion with Penelope, went away again to Etruria and then settled at Gortynaea, where he died greatly respected by the inhabitants; or his remark that Medea was in love with Sisyphus.[33] His story of Cillus, the charioteer of Pelops, was told in recounting the *Ktisis* of Cilla.[34] There is no telling where his famous digressions may have led him, but, considering the generous number of fragments, there is a remarkable scarcity of mythological allusions. Since he is quoted not only by authorities whose main interest is historical, but also by scholiasts and lexicographers who are on the look out for mythological exposition and interpretation, it must be supposed that his work was in fact not very helpful to them in this particular matter. Needless to say, he expressed his view about the date of Homer.[35]

On the other hand, his interest in "the strange and the marvellous" is well illustrated by the fragments. They show that he was interested in foreign peoples and their customs after the manner of Herodotus. Dionysius of Halicarnassus tells us that he included in his history references to "anything remarkable or unusual that each land and sea produced." [36] We find that his digressions took him as far afield as the dwellers by the Ocean, Tartessus, and Paphlagonia; and that in his treatment of the various campaigns of Philip he found opportunities to describe the curious customs of Paeonians, Illyrians, and Scythians.[37]

His love of personalities and anecdotes about historical characters is equally well illustrated.[38] These anecdotes are not confined to the period formally covered by the *Hellenica* and the *Philippica*. In Book XXI of the *Philippica* he pointed out that Peisistratus kept no guards on his estates, but allowed all comers to help themselves freely to his produce—a generous move which

---

[32] F.381—Str. 1.2.35: Θεόπομπος δὲ ἐξομολογεῖται φήσας ὅτι καὶ μύθους ἐν ταῖς ἱστορίαις ἐρεῖ κρεῖττον ἢ ὡς Ἡρόδοτος καὶ Κτησίας καὶ Ἑλλάνικος καὶ οἱ τὰ Ἰνδικὰ συγγράψαντες.

[33] F.354, 356.

[34] F.350.

[35] F.205.

[36] See note 31 above.

[37] F.62, 200, 201, 179; 38, 39, 45.

[38] Cf. F.20 (Lysander), F.22 (Agesilaus), F.31 (Cotys of Thrace).

Cimon imitated; and in Book X there was an excursus on "Athenian demagogues," which included various anecdotes about Cimon and Themistocles: Cimon's generosity and his responsibility for introducing corrupt practices into Athens, and the wealth of Themistocles, which he used in order to bribe the ephors when the walls of Athens were being built.[39]   His anti-Athenian prejudice appears in his attempt to belittle the part played by the Athenians in repelling the Persian invasion.[40]

His interest in the details and origins of Athenian institutions is attested by only one fragment: he explained the origin of the festival known as Χύτροι; he also described the origin of the Carneia at Sparta, and explained who were the Spartan ἐπεύνακτοι and the κατωνακοφόροι at Sicyon.[41]

It appears, then, from the fragments that Theopompus did not cling so closely as Ephorus to the traditional methods and that his work does not stand in such a close relation to the *Atthides* as that of Ephorus.   He has chosen a limited period for treatment in both his works; he is not particularly well disposed towards Athens and is severely critical of some Athenian statesmen of earlier days; his digressions seem to have been concerned for the most part with historical personalities or the strange customs of foreign lands, rather than with Athenian *aetia;* and there is no adequate evidence that he was particularly interested in the details of Athenian religious or mythological tradition.

Entirely different in character from the work either of Ephorus or of Theopompus is Aristotle's *Constitution of Athens*.   Although not formally a history, but a description of the workings of government, in its discussion of the historical development of the constitution it dealt with material which occupied the attention of Attic historians; and Aristotle's treatment of such material is particularly interesting for our purposes, since at least two of the Atthidographers, Cleidemus and Androtion, preceded him.   We have already seen that he probably knew the *Atthis* of Androtion and disagreed with some of its conclusions;[42] no further attempt will be made here to solve the problem of his sources; the present chapter is concerned only with his attitude towards the traditions

[39] F.135, 89, 90, 85, 86.
[40] F.153.
[41] F.347, 357, 171, 176.
[42] See Chap. 4, pp. 82–84 above.

of Attic historiography in general and the extent to which he was interested in its traditional subjects.[43]

It becomes clear at once that, although there are certain obvious points of resemblance, several of the familiar features of an *Atthis* are absent. Aristotle's interest is primarily in political rather than in religious institutions; he describes the functions of the magistrates thoroughly, as some of the Atthidographers evidently did; but religious matters claim his attention only in so far as magistrates are concerned with regulating them. He is, therefore, content to point out that certain preparations are supervised by a certain official, but takes it for granted that his readers are familiar with what follows when the work of the official is done. For example, he points out the duties of the Archon and the Basileus in connection with the Dionysia and other festivals, but is not led on into a digression about the nature or purpose of these celebrations. The following sentence is typical of his manner: "He (the Archon) is in charge of the processions that take place in honour of Asclepius (when the mystae remain indoors) and those at the Great Dionysia, with the assistance of the Epimeletae. These latter were formerly elected by the people, ten in number, and paid the cost of the procession out of their own pockets, but now the people appoints one from each tribe by lot and assigns them 100 minae to cover the expenses. The Archon also supervises the procession at the Thargelia and that in honour of Zeus Soter."[44] The length of the explanatory notes in Sandys's edition shows how much he left out, which an Atthidographer might properly have added.

One reason why Aristotle does not concern himself with the origins of these festivals, which might be regarded as part of the Athenian πολιτεία, is that the discussion would lead him back into mythical times. Though he makes no statement in the surviving portion of the text about his attitude towards myths and the history of very early days in Athens, he is in fact much stricter even than Thucydides in excluding mythological material. His discussion of legendary times was confined to the opening section of

---

[43] It seems reasonable to make Aristotle himself responsible for some characteristics of his work, rather than his supposed source, the mysterious "Anonymus," about whom Otto Seeck ("Quellenstudien zu des Aristotles Verfassungsgeschichte Athens," *Kl* 4 [1904] 164–81, 270–326) and A. von Mess ("Aristoteles 'Aθ. Πολ. und die politische Schriftstellerei Athens," *RhM* 66 [1911] 356–92) speak so confidently. Cf. also P. Cloché, "Hypothèses sur l'une des sources de l' 'Aθ. Πολ.," *MB* 29 (1925) 173–84.

[44] 56.4–5.

the treatise, which is lost.   In that part of his work he pointed out
the divine origin of Ion and described the constitution set up by the
early settlers under his leadership.[45]   He also apparently described
the constitution set up in the time of Theseus and gave a fairly
detailed account of his political reforms.[46]   But the fragments mention no other event in Theseus' life beyond his journey to Scyros
and his death there at the hands of Lycomedes;[47] Aristotle is
never cited as an authority for any of the heroic ἆθλα of Theseus.
Apart from this treatment of τὰ πάνυ παλαιά there is no allusion to a
mythological character or incident in the rest of his work.   This
absence of mythological discussion distinguishes the *Constitution of
Athens* very sharply from the work of the Atthidographers.   Just
because of his strict attitude in this respect, it would be particularly
interesting to know exactly how much space in the lost portion of
his treatise Aristotle allotted to the great Athenian hero of the
Atthid tradition, and how much importance he attached in general
to traditions about such remote times.

On the other hand, Aristotle does discuss the origin of some of
the political offices, more particularly if their origin is connected
with any of the turning points in Athenian constitutional history.
The famous landmarks in Attic history, the few significant incidents
known to have taken place in the otherwise obscure period of the
seventh and sixth centuries, naturally take on an added significance
in his work, because they mark stages in the development of the
constitution.   One might expect him, in strictness, to confine himself to constitutional aspects of the work of Peisistratus and Solon
and to disregard the anecdotes which embellished the history of
their times.   This, however, is not the case.   He is sufficiently

---

[45] Harp. *s.v.* Ἀπόλλων πατρῷος and Sch. Ar. *Av.* 1527—Arist. *Resp. Ath.* (Oxford
text, ed. Kenyon), fg. 1: πατρῷον τιμῶσιν Ἀπόλλωνα Ἀθηναῖοι, ἐπεὶ Ἴων ὁ πολέμαρχος
Ἀθηναίων ἐξ Ἀπόλλωνος καὶ Κρεούσης τῆς Ξούθου <γυναικὸς> ἐγένετο. Epit. Heraclid.
1: Ἀθηναῖοι τὸ μὲν ἐξ ἀρχῆς ἐχρῶντο βασιλείᾳ, συνοικήσαντος δὲ Ἴωνος αὐτοὺς τότε πρῶτον
Ἴωνες ἐκλήθησαν. Note also the reference in *Resp. Ath.* 41.2 to the πρώτη . . .
μετάστασις τῶν ἐξ ἀρχῆς, Ἴωνος καὶ τῶν μετ' αὐτοῦ συνοικησάντων· τότε γὰρ πρῶτον εἰς
τὰς τέτταρας συνενεμήθησαν φυλὰς καὶ τοὺς φυλοβασιλέας κατέστησαν.

[46] Plu. *Thes.* 25—Kenyon, fg. 2 (see note 45). Cf. *Resp. Ath.* 41.2: δευτέρα δὲ καὶ
πρώτη μετὰ ταύτην ἔχουσα πολιτείας τάξιν ἡ ἐπὶ Θησέως γενομένη, μικρὸν παρεγκλίνουσα
τῆς βασιλικῆς.

[47] Sch. Vat. Eur. *Hipp.* 11: Ἀ. ἱστορεῖ ὅτι ἐλθὼν Θησεὺς εἰς Σκῦρον ἐπὶ κατασκοπὴν
εἰκότως διὰ τὴν Αἰγέως συγγένειαν ἐτελεύτησεν ὠσθεὶς κατὰ πετρῶν, φοβηθέντος τοῦ Λυκο-
μήδους τοῦ βασιλεύοντος. Note the εἰκότως, which argues a rationalistic approach to
the myth—what R. W. Macan calls "an *a priori* method in historical research"
("Ἀθηναίων Πολιτεία," *JHS* 12 [1891] 39).

interested in the character of Solon to mention and reject the story told about him by his detractors: that he joined with the γνώριμοι in buying land on borrowed money, which, when all debts were cancelled by the Seisachtheia, they were not obliged to repay. He is also interested in the character of Peisistratus, whom he regards as a kindly, tolerant man, ready to lend money to those who needed it in order to keep their farms in operation; and he records, without comment or reference to Herodotus, the story that his second exile was the result of his marital disagreement with the daughter of Megacles.[48]

There are other examples of his interest in anecdotes and personal details about Athenian public men. He makes some observations on the characters of Themistocles, Aristeides, and Ephialtes.[49] His remarks about the generosity of Cimon seem to indicate some regard for the tradition recorded by Theopompus.[50] He also remarks on the rude manner of Cleon in haranguing the people, and how Cleophon came to the assembly drunk and wearing his breastplate in order to make a truculent speech denouncing any effort to make peace with Sparta.[51] His recording of such trifles as these shows not only his lack of any new significant information about these men, but the influence on him of the anecdotal type of history; his discussion of fifth century Athenian statesmen and the part they played in Athenian constitutional development is not as illuminating as the accounts of Herodotus and Thucydides.[52]

Furthermore, though he mentions Herodotus only once, Thucydides, Ephorus, and Theopompus not at all, he is ready enough to take part in controversy. Sometimes the matter of controversy is trifling—as, for example, his uncertainty about the native place of the woman disguised as Athena who helped in the restoration of Peisistratus.[53] He dismisses as "obvious nonsense" the story that Peisistratus was enamoured of Solon, because it is chronologically impossible;[54] but his story about the intrigue of Themistocles and

---

[48] 6.2–3; 16.2; 15.1 (cf. Hdt. 1.60–61). Note also his treatment of the story of Harmodius and Aristogeiton in 18, which is in the nature of a digression.

[49] 23.3; 25.1.

[50] 27.3; Theopomp. F.89, 90. See above p. 99.

[51] 28.3; 34.1.

[52] Yet one need not go so far as to follow Seeck (*op. cit.* 287) in believing that Aristotle ignored Herodotus, Thucydides, and Xenophon, except in so far as his source, the political pamphleteer "Anonymus," quoted or copied from them.

[53] 14.4. Cf. Cleidemus fg. 24; and chap. 4 p. 68 above.

[54] 17.2.

Ephialtes, which resulted in the weakening of the Areopagus, appears equally impossible on the same grounds.[55] More serious points of controversy concern the date of the establishment of the Archon's office, the question of the census of the Knights, and the merits of Theramenes.[56] In none of these instances does he mention the names of the conflicting authorities.

A most important point is his attention to chronological detail. In indicating dates he is much more careful and complete than any of the earlier historians, and he indicates the year as a rule by reference to the archon's name. Frequently also he will date an event in relation to an earlier event. Thus Aristeides is said to establish the tribute "in the third year after Salamis when Timosthenes was archon," and the generals in command at Arginusae are condemned "in the seventh year after the fall of the Four Hundred, when Callias was archon." [57] Not only are the dates of events in these comparatively recent times thus definitely set down, but indications for earlier times are by no means lacking, not only for Peisistratus, but even for the disturbances after Solon's reforms.[58] The institution of the Archon (subsequent to the Basileus and Polemarch) is said to be in the year either or Medon or of Acastus, the legislation of Draco in the archonship of Aristaechmus.[59]

Naturally this generosity in the indication of dates raises the question of Aristotle's sources. The actual correctness of the indications is not a point which concerns us here. The significant thing is that he gives the date, as a rule, without any discussion or argument, and without revealing whence his information comes. To a certain extent he is indebted to documentary sources—as, for example, when he mentions the actual proposer of a law, which he does on a few occasions. His accounts of the revolutions of the Four Hundred and of the Thirty contain several such allusions,[60] and here he certainly had access to official documents. In a different category, however, is his mention of Aristion as the man who proposed that Peisistratus should have a bodyguard.[61] A

[55] 25.3. For discussion see Sandys's note and the works cited there. Cf. also A. von Mess, op. cit. 389–90.

[56] 3.3; 7.4; 28.5.

[57] 23.5; 34.1 (it should be the sixth, not the seventh year; see Sandys's note).

[58] 13.1.

[59] 3.3; 4.1.

[60] 29.1; 32.1; 34.3; 40.2.

[61] 14.1.

detail of this kind about an event of the sixth century is naturally suspect and suggests a literary source—in other words, the invention of a predecessor; unless a forger of documents is to be blamed for supplying dates, as well as for forging the constitution of Draco.[62]

Students of Aristotle may think that they have achieved some purpose if their statement that he borrowed from Critias or from some other writer is not susceptible of disproof. It is always easy to settle the matter of an historian's source by naming a writer of whom comparatively little is known. In reality, to say that Aristotle borrowed material from an earlier writer is only to raise another question: How did this earlier writer obtain it? If it could be proved that the earlier writer invented it, then something would be achieved; but this result can never be established. It is true that by his readiness to raise controversial issues and mention divergent opinions Aristotle shows he did not ignore those who had preceded him in writing Athenian history and discussing Athenian institutions. But in refusing to mention the names of his predecessors he conforms to a tradition familiar from the time of Herodotus.

Aristotle's readiness to date events more than two centuries before his time without explaining whence his knowledge came suggests that some kind of chronological system had been established when he wrote. It is the Atthidographers, more than any others, who have been suspected of performing this pioneer work,[63] valuable or misleading as the case may be. One of our tasks in the two chapters which follow will be to investigate the grounds for this suspicion.

[62] See chap. 1 p. 23 above.

[63] Wilamowitz held (*Aristoteles u. Athen, passim*) that an *Atthis* written in the fifth century had established dates for historical events since Solon's time and that Aristotle derived his dates ultimately from that source. This view has been questioned often (most recently by W. Kolbe, "Diodors Wert für die Geschichte der Pentekontaetie," *H* 72 [1937] 241–69), and the actual fragments of the Atthidographers offer very little evidence in support of it.

CHAPTER VI

Philochorus and Ister

I. PHILOCHORUS

The fragments of Philochorus in Müller's collection are much more numerous than those of the other Atthidographers, and their number has been very considerably increased since his time.[1] The most important additions are the quotations from his *Atthis* given by Didymus, nicknamed Χαλκέντερος, in his commentary on the speeches of Demosthenes, a portion of which is preserved on a papyrus acquired by the Berlin museum authorities in 1901. Didymus, who lived in the Ciceronian period, evidently used the *Atthis* as a convenient work of reference for the history of Demosthenes' time. This papyrus thus provides us with valuable information about the treatment of the fourth century by Philochorus. His treatment of the fifth century is best illustrated by the quotations which are given in the scholia on Aristophanes. Wilhelm Meiners,[2] who wrote before the papyrus had been published, argued that these scholia were to a great extent derived from Didymus, and the familiarity of Didymus with the work of Philochorus which the papyrus reveals naturally strengthens his case considerably. On the other hand, the hypothesis that Philochorus owes his frequent mention by later authorities to citations by Ister in his Ἀτθίδων Συναγωγή [3] rests on no sure foundation. For the present purpose, however, it is of little importance from what quarter these later writers obtained their information. They cited Philochorus by name and frequently quoted his actual words; in this manner they bear witness to the reputation which he enjoyed.

His name appears in two Byzantine library catalogues of the sixteenth century,[4] so that there is ground for believing that his works survived much longer than those of the other Atthidographers. This does not mean, however, that Natale Conti, who

---

[1] For references see the bibliography.

[2] *Diss. Philol. Halenses* 11.219–402.

[3] Cf. G. Gilbert, *Ph* 33 (1874) 46–54. See p. 137 below.

[4] Cf. K. Krumbacher, *Gesch. der byzantin. Lit.* 508–09, who gives references; he considers that no faith can be put in the entries.

frequently refers to Philochorus in his *Mythologia*,[5] had ever seen a text of this author. Conti certainly makes a formidable display of classical knowledge; but it has been shown that most of his references to Philochorus are derived from scholiasts and lexicographers; and that in a few other cases he attached the name of Philochorus to a story for which no other equally interesting authority was attested.[6] Freculphus, bishop of Lisieux in the ninth century, also mentions Philochorus, but he seems to have quoted only at second hand, since his citations correspond with what is recorded by Eusebius and Syncellus.[7]

Apart from these citations in an unexpected quarter, the chief authorities for the fragments are, as usual, the lexicographers and scholiasts, particularly the scholiast on Aristophanes, who frequently cites Philochorus in preference to Thucydides for details in the history of the Pentecontaetia and the Peloponnesian War; and there are a few references in the patristic writers and in Athenaeus.

Just enough is recorded of the life of Philochorus to give us some idea of the position he occupied in Athens. Suidas tells us that he was a "prophet and a reader of the signs of sacrifice," [8] as indeed his works *On Prophecy* and *On Sacrifices* would lead us to believe.

[5] Müller gives only three references: Fg. 175, 29, 174—1.36, 3.249, 9.1020 (these references are to the third edition of 1619; the *Mythologia* first appeared in 1551).

[6] His few references to Phanodemus should certainly be explained in the same way (see Müller on Phanodemus fg. 3a). For a discussion of the whole question see R. Dorschel, *Qualem in usurpandis veterum scriptorum testimoniis Natalis Comes praestiterit fidem* (Diss. Greifswald, 1862).

[7] Roersch, *MB* 1 (1897) 146–49, compares six passages from Freculphus' *Chronica* (Migne, *Patrol. Lat.* 106.948, 956, 957, 959, 963, 969) with Fg.10, 28, 30, 23, 39, 53.

[8] The complete text of Suidas (in Adler's edition) is as follows: Φιλόχορος, Κύκνου, Ἀθηναῖος, μάντις καὶ ἱεροσκόπος· γυνὴ δὲ ἦν αὐτῷ Ἀρχεστράτη. κατὰ δὲ τοὺς χρόνους γέγονεν ὁ Φιλόχορος Ἐρατοσθένους, ὡς ἐπιβαλεῖν πρεσβύτῃ νέον ὄντα Ἐρατοσθένει. ἐτελεύτησε δὲ ἐνεδρευθεὶς ὑπὸ Ἀντιγόνου, ὅτι διεβλήθη προσκεκλικέναι τῇ Πτολεμαίου βασιλείᾳ. ἔγραψεν Ἀτθίδος βιβλία ιζ'· περιέχει δὲ τὰς Ἀθηναίων πράξεις καὶ βασιλεῖς καὶ ἄρχοντας, ἕως Ἀντιόχου τοῦ τελευταίου τοῦ προσαγορευθέντος θεοῦ· ἔστι δὲ πρὸς Δήμωνα· Περὶ μαντικῆς δ', Περὶ θυσιῶν α', Περὶ τῆς Τετραπόλεως, Σαλαμῖνος κτίσιν, Ἐπιγράμματα Ἀττικά, Περὶ τῶν Ἀθήνησιν ἀγώνων βιβλία ιζ', Περὶ τῶν Ἀθήνησιν ἀρξάντων ἀπὸ Σωκρατίδου καὶ μέχρι Ἀπολλοδώρου, Ὀλυμπιάδας ἐν βιβλίοις β', Πρὸς τὴν Δήμωνος Ἀτθίδα, Ἐπιτομὴν τῆς ἰδίας Ἀτθίδος, Ἐπιτομὴν τῆς Διονυσίου πραγματείας περὶ ἱερῶν, Περὶ τῶν Σοφοκλέους μύθων βιβλία ε', Περὶ Εὐριπίδου, Περὶ Ἀλκμᾶνος, Περὶ μυστηρίων τῶν Ἀθήνησι, Συναγωγὴν ἡρωίδων ἤτοι Πυθαγορείων γυναικῶν, Δηλιακὰ βιβλία β', Περὶ εὑρημάτων, Περὶ καθαρμῶν, Περὶ συμβόλων. The statement that Philochorus was young when Eratosthenes (born 275) was an old man contradicts the rest of the evidence. Possibly Suidas has made a mistake; but the emendation Ἐρατοσθένῃ for Ἐρατοσθένει (suggested by Siebelis) would make Philochorus an old man in the youth of Eratosthenes and remove all difficulty.

Proclus is the only authority who actually calls him an Exegetes,[9] but a quotation by Dionysius of Halicarnassus from the *Atthis* establishes the fact beyond any doubt.[10] Dionysius first quotes from Book VIII his account of events in the archonship of Anaxicrates (307–6 B.C.), when, after the entry of Demetrius Poliorcetes into Athens, the rule of Demetrius of Phalerum was overthrown and a number of people were forced to go into exile. He quotes this passage to show the circumstances under which the orator Deinarchus had to leave Athens; then, to show the circumstances of his return fifteen years later, he goes on:

> And in Book IX Philochorus writes: "With the end of this year and the beginning of the next, a sign was observed on the Acropolis as follows. A dog entered the temple of Athena Polias, penetrated as far as the *Pandroseion*, and after climbing up onto the altar of *Zeus Herkeios* beneath the olive tree lay down there (and according to Athenian tradition no dog should go up on the Acropolis). At the same time there was also a sign observed in the sky. During the day time, when the sun was out and the sky was clear, a star could be seen plainly for some time. We were questioned about the meaning of this sign and this strange phenomenon, and we said that both were signs indicating a return of exiles—not that this would involve a revolution, but it would take place without any disturbance of the established order; and our reply was found to be correct."

This description is clear evidence that Philochorus followed the calling of an Exegetes in Athens and was old enough in the year 292 to be consulted about the meaning of signs and portents.[11] Suidas says that his *Atthis*, in seventeen books, extended down to the time of Antiochus Theos, and adds that he was killed "at the order of Antigonus because he was said to have favoured the cause of Ptolemy." The reign of Antiochus Theos begins in 262–1, and Athens was captured by Antigonus Gonatas in 263–2; evidently, therefore, Philochorus was connected with the group of intellectuals who had planned the Chremonidean War, in response to an invitation from Egypt, and he was subsequently (perhaps not until some

[9] Fg. 183—Ad Hes. *Op.* 808 (Gaisford, *Poetae Graeci Minores* 2.441).

[10] *Dein.* 3—Fg. 144, 146.

[11] Cf. Wilamowitz, *Philol. Unters.* 4.204; W. S. Ferguson, *Hellenistic Athens* 140–41. There are difficulties in supposing that Book IX went down as late as 292. Either the number is incorrectly quoted or else the passage occurs in a digression. See pp. 111–12 for discussion of the chronological arrangement of the *Atthis*. Müller contradicts himself, saying that Philochorus must have been old enough to be an Exegetes in 306 (since he thinks the passage refers to this year), but at the same time suggesting 320 as the date of his birth (*FHG* 1.lxxxiv).

years later) executed for his complicity in the affair.[12]   If he was old enough to be an Exegetes in 292 and still young enough in 267–62 to be politically active, Müller's suggested dates of 320–260 for his lifetime are reasonable enough; and the Κύκνος Φιλοχόρου 'Αναφλύστιος mentioned in an Attic inscription as receiving a crown in 334–3 may well be his father.[13]

Philochorus wrote many other works besides his *Atthis*. Twenty-four titles are attested altogether, some of them mentioned only by Suidas.[14] Some of these titles may be alternative names for the *Atthis* or some portion of it; for example, the *Reply to Demon* [15] and *On the Athenian Archons from Socratides to Apollodorus* (373–318 B.C.).[16] Equally uncertain are the titles *On the Tetrapolis* and *On the* 'Αγῶνες *at Athens*.   There are three citations from the former work, two of which could be referred equally well to an *Atthis*. Athenaeus refers to it for a discussion of the παράσιτοι in the cult of Heracles at Athens, shortly after quoting the *Atthis* of Cleidemus for the same point; [17] and Suidas cites it together with the *Atthis* of Ister for the name Τιτανὶς γῆ as a name of Attica.[18] We have already seen that discussion of Athenian religious customs and topographical points was characteristic of *Atthides*. The work on the 'Αγῶνες is mentioned only by Suidas and there are no citations from it; and of the two fragments assigned to it by Müller, one would certainly be appropriate to an *Atthis:* a passage quoted *verbatim*

[12] Cf. D.L. 7.24; W. W. Tarn, *CAH* 7.220, 706, 712; Ferguson, *Hellenistic Athens* 188.

[13] *IG* 2².1750.   Cf. Wilamowitz, *H* 20 (1885) 631.

[14] With the following discussion cf. Roersch, *MB* 1 (1897) 137–157.

[15] Suidas says the *Atthis* was written πρὸς Δήμωνα, but later in his list he gives a separate title πρὸς τὴν Δήμωνος 'Ατθίδα; and Harpocration refers to ἡ πρὸς Δήμωνα ἀντιγραφή (Fg. 115).

[16] Böckh (*Kl. Schr.* 5.401–12) thinks this may be an earlier and less pretentious work.   For the title, cf. the 'Αρχόντων ἀναγραφή of Demetrius of Phalerum (*FGrH* 2B, no. 228).   Note also the two books of *Olympiads* which Suidas attributes to Philochorus.   He also credits him with an *Epitome* of his own *Atthis*, but elsewhere attributes the work to Pollio of Tralles (*s.v.* Πωλίων, ὁ 'Ασίνιος χρηματίσας . . . ἔγραψεν ἐπιτομὴν τῆς Φιλοχόρου 'Ατθίδος).

[17] Ath. 6.235 A–D—Cleidemus Fg. 11, Philoch. Fg. 156.

[18] Suid. *s.v.* Τιτανίδα γῆν· οἱ μὲν τὴν πᾶσαν, οἱ δὲ τὴν 'Αττικήν. ἀπὸ Τιτηνίου, ἑνὸς τῶν Τιτάνων ἀρχαιοτέρου, οἰκήσαντος περὶ Μαραθῶνα, ὃς μόνος οὐκ ἐστράτευσεν ἐπὶ τοὺς θεούς, ὡς Φιλόχορος ἐν Τετραπόλει, "Ιστρος δ' ἐν α' 'Αττικῶν. Τιτᾶνας βοᾶν, ἐβοήθουν γὰρ τοῖς ἀνθρώποις ἐπακούοντες, ὡς Νίκανδρος ἐν α' Αἰτωλικῶν. ἐνομίζοντο δὲ τῶν Πριαπωδῶν θεῶν εἶναι. This is the text as given in Adler's edition of Suidas. Müller's printing of Philoch. Fg. 157 and Ister Fg. 1 is misleading. The third reference to the *Tetrapolis*, in the scholia on Sophocles (Fg. 158), is concerned with details of religious procedure.

by Athenaeus (without reference to any particular book), which describes how the Athenian audience at the Dionysia was served with wine and cakes whilst sitting in the theatre.[19] Furthermore, since Suidas says that there were seventeen books in the Ἀγῶνες—the same number as in the *Atthis*—it seems quite probable that he is merely confusing matters in his usual manner and that there was no separate work on *Contests*.

These conclusions are of course by no means certain. It is quite possible that Philochorus did write independent works under the titles quoted, and no particular purpose is served by attempting to reduce the quantity of his literary output. The real object of the foregoing argument is to show that in these works, whether they are separate from the *Atthis* or not, Philochorus dealt with material similar to that which occupied him in the *Atthis*. Any fragments, therefore, which are or could be attributed to these works, must be taken into account in discussing the scope of his historical work and his interests as an antiquary. It is unfortunate that no fragments are quoted from his work on Attic inscriptions (Ἐπιγράμματα Ἀττικά). Böckh, himself the founder of modern Attic epigraphical study, thought that the existence of such a work spoke well for the reliability of Philochorus as an historian.[20]

The other works fall readily into three separate categories. First, there are works devoted to regions outside of Attica. To the *Deliaca*, mentioned by Suidas, Müller assigns two fragments about the island of Tenos,[21] concerning the cult of Poseidon there and statues of Poseidon and Amphitrite from the workshop of an Athenian sculptor. The *Epirotica* suggests possible connections of Philochorus with Pyrrhus. Here also Müller offers two fragments:[22] one simply concerns the name Ellopia, applied to the district round Dodona as it was to Euboea; the other mentions the name of the city Boucheta in Epirus. No doubt it was the religious customs of Delos and Epirus, with their famous sanctuaries, which occupied the attention of Philochorus rather than their history or topography. His *Deliaca* and *Epirotica* were probably not unlike the *Deliaca* and

---

[19] Ath. 11 464F—Fg. 159. Müller also assigns Fg. 160 to this work; it might equally well be referred to the Περὶ Τραγῳδιῶν, since it is an anecdote about the actor Polus.

[20] *Kl. Schr.* 5.399f.

[21] Fg. 184, 185.

[22] Fg. 186, 187. The authenticity of the latter frag. is doubtful; Suidas cites Philochorus, but Harpocration refers to "Philostephanus in the *Epirotica*." Cf. Stiehle, *Ph* 4 (1849) 391; 8 (1853) 639.

Iciaca of Phanodemus.[23]   The *Founding of Salamis* is known only from Suidas' list.

There are a number of works on religious topics.   Little need be said about the treatises *On Festivals, On Sacrifices,* and *On Prophecy,*[24] except that Philochorus was not the only Atthidographer to write such treatises.   Somewhat different is the work *On Days,* apparently a modernized and systematic collection of the kind of folklore which Hesiod used in his *Works and Days,* including remarks on the appropriate sacrifices for each day of the month and year.   Müller quotes eight fragments from this work, and Reitzenstein has added several more from the Lexicon of Photius and other sources.[25]   Nothing is known of the other religious works mentioned in Suidas' list: *On Expiations, On the Mysteries at Athens, An Epitome of the Treatise of Dionysius on Holy Things;* Περὶ Συμβόλων, the last item on his list, suggests either a discussion of contracts or a work on the interpretation of omens.

There remain his works of literary criticism: *On Euripides,* which is known through several fragments;[26] the *Letter to Asclepiades,* which is possibly the same as *On Tragedies;*[27] *On the Myths of Sophocles* and *On Alcman,* which are known only from Suidas' list.   These works evidently belong to quite a different category from his historical, antiquarian, and priestly studies.   They may be looked upon as forerunners of the many literary studies written in Alexandrian times; Demetrius of Phalerum, with whom Philochorus seems to have had something in common,[28] also wrote treatises on the *Iliad* and the *Odyssey.*   It seems likely that *On Inventions* and *A Collection* (Συναγωγή) *of Heroines or Pythagorean Women* were handbooks or compilations for the literary student; but nothing is known of these works except their titles, which are listed by Suidas.[29]

We know far more about the *Atthis* than about the other works of Philochorus.   Müller has over a hundred and fifty fragments

---

[23] Cf. Chap. 4, p. 72 above.

[24] Fg. 161, 163, 173–75, 190–93.

[25] Fg. 176–83; Reitzenstein, *NGG,* phil.-hist. Kl., 1906, 40–48.

[26] Fg. 165–69.

[27] The two works are cited for the same material. Cf. Sch. Eur. *Hec.* 1—*FHG* 4.648 and Stiehle, *Ph* 8 (1853) 640. Ἄλυπος in Photius is evidently a mistake for Ἀσκληπιάδης.

[28] Cf. the list of Demetrius' works in D.L. 5.80–81.

[29] Müller somewhat diffidently assigns Fg. 188, 189 (about Linus) to the Περὶ Εὑρημάτων.

which should certainly be attributed to it, and with the ten new citations from Didymus and the other fragments added by Stiehle and Strenge, the evidence for this *Atthis* is much fuller than for any other work of its kind. Thanks to the number of the fragments, we can establish with reasonable certainty how much ground he covered in each of the first ten books. Böckh's conclusions on this point, based on the collection of Lenz and Siebelis, have not been upset by the new fragments and his essay is still valuable. The fragments in Müller's collection are arranged in accordance with these conclusions, which may conveniently be summarized here. References to book numbers are fairly frequent in the fragments, so that detailed argumentation is not necessary.

Book I did not go beyond King Cecrops. Book II dealt with Cecrops and his royal successors and with the story of Theseus; the exact point at which the book ends cannot be established; Böckh thinks it may be the archonship of Creon, the first archon, whose traditional date is 683–2 B.C. Book III dealt with Solon, as a reference to "the oath over the stone" shows (Fg. 65), and with the Peisistratids (Fg. 69); references to the names of demes (Fg. 71–76) and to the procedure of ostracism (Fg. 79b) suggest the legislation of Cleisthenes; whilst the *Theorikon* (Fg. 85) seems to belong to the age of Pericles, and the Laconian city of Aethaea (Fg. 86) would most naturally be mentioned in connection with the Helot revolt.[30] The dividing line between Books III and IV must come somewhere in the middle of the fifth century. Book IV seems to have reached the end of the Peloponnesian War; it may even have gone as far as 392, the point where Theopompus ended his *Hellenica*. Book V is cited for the symmories, first established in 377 (Fg. 126), and Book VI for the group of 1200 wealthy people selected by the second law about symmories, as proposed by Periander in 357 (Fg. 129); hence Philip's accession to the throne of Macedonia (359) seems a likely terminating point for Book V. Book VI is quoted for various events in the war with Philip,[31] and since no event before the rule of Demetrius of Phalerum in Athens is mentioned in any fragment from Book VII, Böckh thinks the dividing line may be the year in which his work *On the Archons* ended—318.

[30] Cf. Th. 1.101.

[31] A citation of the sixth book of the *Atthis* in the *Academicorum Philosophorum Index Herculensis*, ed. S. Mekler, col. 2.5–6, evidently refers to an incident in Plato's old age.

The later books will then be on a quite different scale. Book VII dealt with the rule of Demetrius of Phalerum and his reforms, and the submission of Athens to Demetrius Poliorcetes was related in Book VIII (Fg. 144). A remark about the irregular initiation of Demetrius Poliorcetes into the mysteries at Athens, in the year 302, is referred to Book X;[32] but the passage quoted by Dionysius of Halicarnassus from Book IX, in which Philochorus described his prophecy of the return of the exiles, must refer to 292.[33] It seems that either Harpocration or Dionysius has made a mistake about the book number, unless one of the passages occurred in a digression; but in any case the dividing line between Books IX and X cannot be established. For the end of Book X and the division of the material from 292 to 261 among the last seven books no evidence is available. Laqueur points out that the Alexandrian scholars were not so much interested in Attic history of the third century as in the period of the famous orators; and he thinks that the lack of citations from the later books of the *Atthis* may be due to this cause.[34]

With the general outline of the *Atthis* thus established, it will now be possible to discuss more in detail the characteristics of his work as they are revealed in the fragments.

His interest in religious questions is not confined to the early books; it is best illustrated by fragments from Books I and II, but there are signs of it in all parts of the *Atthis*. Many fragments refer to his discussion of Athenian religious rites and their origin: how Amphictyon instituted the worship of the nymphs as daughters of Dionysus and how Erichthonius introduced the custom that girls should carry baskets and old men olive branches in the Panathenaic procession;[35] how the worship of Hermes ὁ πρὸς τῇ πυλίδι began when the Athenians started to fortify the Peiraeus, and the cult of Hermes Ἀγοραῖος also dated from the fifth century.[36] He also spoke of the

---

[32] Fg. 148—Harp. *s.v.* ἀνεπόπτευτος· . . . ὁ μὴ ἐποπτεύσας. τί δὲ τὸ ἐποπτεῦσαι, δηλοῖ Φιλόχορος ἐν τῇ δεκάτῃ· "Τὰ ἱερὰ οὗτος ἀδικεῖ πάντα τά τε μυστικὰ καὶ τὰ ἐποπτικά." καὶ πάλιν· "Δημητρίῳ μὲν οὖν ἴδιόν τι ἐγένετο παρὰ τοὺς ἄλλους, τὸ μόνον ἅμα μυηθῆναι καὶ ἐποπτεῦσαι, καὶ τοὺς χρόνους τῆς τελετῆς τοὺς πατρίους μετακινηθῆναι." For the date of this incident see Ferguson, *Hellenistic Athens* 122.

[33] See p. 107 above.

[34] *RE* 19.2436.

[35] Fg. 18, 25, 26.

[36] Fg. 80, 81, 82—Hsch. *s.v.* Ἀγοραῖος· Ἑρμῆς οὕτως ἐλέγετο ὄντως, καὶ ἀφίδρυτο Κεβρίδος ἄρξαντος, ὡς μαρτυρεῖ Φιλόχορος ἐν τρίτῳ. Böckh, *Kl. Schr.* 5.411, thought Κεβρίδος must be a mistake, perhaps for Hybrilides, archon 491–0. But Cebris may have been archon in 486–5, for which year no other name is attested. Cf. Wilamowitz, *H* 21 (1886) 600, and Sandys's note on Arist. *Resp. Ath.* 22.5.

Χύτροι and the competitions held during the festival; and of the less known festivals of the Γενέσια and the Δειπνοφόρια, the latter of which was instituted in the time of Theseus.[37]   Apart from purely Attic cults and festivals, we find that he described the worship of the half-male Aphrodite in Cyprus and that he had something to say about the qualities of Dionysus as a divine character.[38]

His remarks about Dionysus throw light on his general attitude towards the traditional Athenian religion. "We must not think of Dionysus," he wrote, "as a kind of buffoon and a disreputable clown." [39]   He insisted that the god's grave really did exist at Delphi,[40] and that he should be regarded as a soldierly general, not an effeminate drunkard; that the ground for misrepresenting him in female form was because his army had included women as well as men.[41]   Remarks of this kind suggest that Philochorus, as an official representative of the traditional religion, was seeking to justify it against the philosophical heretics who complained that it involved respect for barbaric heroes and immoral gods.   It seems that he tried to humanize the old myths, to present the traditional religion in a manner that would appeal to people who had been taught by the sophists and philosophers to look for an ethical basis in religious belief.   Other fragments give further hints of this tendency on his part.   Since Cecrops was supposed to be a "product of the soil" (αὐτόχθων), traditional legend had represented him as half man, half snake; Philochorus insisted that he was called "of double nature" (διφύης) for quite a different reason: either because of his exceptional tallness or because he was an Egyptian and could speak Egyptian as well as Greek.[42]   In a similar way he explained the chthonic mystery of Triptolemus: it was a ship, not a winged

[37] Fg. 137, 164, 164a (FHG 4.648).

[38] Fg. 15; 22–24

[39] Fg. 24—Harp. s.v. Κοβαλεία· . . . κοβαλεία ἐλέγετο ἡ προσποιητὴ μετὰ ἀπάτης παιδία, καὶ κόβαλος ὁ ταύτῃ χρώμενος. ἔοικε δὲ συνώνυμον τῷ βωμολόχῳ. Φιλόχορος ἐν δευτέρῳ ᾿Ατθίδος· "Οὐ γὰρ, ὥσπερ ἔνιοι λέγουσι, βωμολόχον τινὰ καὶ κόβαλον γίνεσθαι νομιστέον τὸν Διόνυσον."

[40] Fg. 22.

[41] Fg. 23—Syncellus 307, ed. Dindorf: βάθρον δέ τι νομίζεται τοῖς ἀγνοοῦσιν ὁ Διονύσου τάφος, στρατηγὸς δὲ δοκεῖ γενέσθαι, καὶ οὕτω γράφεται θηλύμορφος διά τε ἄλλας αἰσχρὰς αἰτίας καὶ διὰ τὸ μιξόθηλυν στρατὸν ὁπλίζειν. ὥπλιζε γὰρ σὺν τοῖς ἄρρεσι τὰς θηλείας, ὥς φησιν ὁ Φιλόχορος ἐν δευτέρῳ.

[42] Fg. 10.   Note also that according to Fg. 13 Cecrops founded the worship of Uranus and Ge in Athens; is this also an explanation of the tales about his earthy origin?

serpent, on which he had travelled when bringing the gift of grain to men.[43]

It is not surprising, therefore, that, after the manner of Plato, he objected to "the many lies that poets tell." [44] The fragments give several examples which show how he revised some familiar legends. For example, the contradiction between Minos, the just law-giver, and Minos, the cruel tyrant, was an obvious stumbling block to the faithful. Philochorus evaded the difficulty, not by the old device of making them separate individuals, but by insisting that the Minotaur was merely a general of Minos called Taurus, "of a cruel and savage disposition"; that at the games given in honour of the dead Androgeos Athenian boys and girls were offered as prizes (appropriately enough, since Athenians were responsible for his death); that Taurus, who was unpopular both with the king and with everyone else, seemed likely to carry off all the prizes, until Theseus threw him in a wrestling bout and so saved the young Athenians from slavery, to the satisfaction of everyone, including Minos himself.[45]

There were other tales which also needed change, if Theseus was to be presented as free from fault. We do not know how Philochorus explained away his faithlessness in deserting Ariadne; Plutarch records only his story of their first meeting: how women were permitted by Cretan custom to watch the games and Ariadne there fell in love with him at first sight.[46] But he could not accept the tale which represented Theseus as fighting with the gods of the lower world. His version is that Persephone is carried off by a king of the Molossians named Aedoneus, who has a gigantic dog named Cerberus; and when Theseus joins Peirithous in an attempt to

[43] Fg. 28. Philochorus remarked that the ship was wrongly interpreted as a winged serpent, ἔχειν δέ τι καὶ τοῦ σχήματος.

[44] Fg. 1—Sch. Pl. *Just.* 374a (Greene, *Scholia Platonica* 402): παροιμία, ὅτι πολλὰ ψεύδονται ἀοιδοί, ἐπὶ τῶν κέρδους ἕνεκα καὶ ψυχαγωγίας ψευδῆ λεγόντων. φασὶ γὰρ τοὺς ποιητὰς πάλαι λέγοντας τἀληθῆ, ἄθλων ὕστερον αὐτοῖς ἐν τοῖς ἀγῶσι τιθεμένων, ψευδῆ καὶ πεπλασμένα λέγειν αἱρεῖσθαι, ἵνα διὰ τούτων ψυχαγωγοῦντες τοὺς ἀκρωμένους τῶν ἄθλων τυγχάνωσιν. ἐμνήσθη ταύτης καὶ Φιλόχορος ἐν Ἀτθίδος αʹ καὶ Σόλων Ἐλεγείαις καὶ Πλάτων ἐνταῦθα. This recalls Plato's criticism of myths in *R.* 377d: ἐάν τις μὴ καλῶς ψεύδηται. No doubt Philochorus had not forgotten what the Muses said to Hesiod:

ἴδμεν ψεύδεα πολλὰ λέγειν ἐτύμοισιν ὅμοια,
ἴδμεν δ', εὖτ' ἐθέλωμεν, ἀληθέα γηρύσασθαι
(*Th.* 27–28).

[45] Fg. 38–40. Demon offered a similar version of the tale, except that Theseus' victory, according to him, was won in a battle in the harbour of Cnossus (Fg. 3).

[46] Fg. 40—Plu. *Thes.* 19.

rescue her, the dog kills Peirithous and Theseus is kept prisoner till Heracles persuades Aedoneus to release him. This escape from deadly peril is then said to have been misconstrued as "a return from the house of Hades"; [47] and instead of an impious act of rebellion against the gods, this adventure of Theseus is presented as part of his mission to civilize the world. This mission is supposed to begin in early youth; he is a very young man when he overpowers the bull of Marathon, and the old country-woman Hecale speaks to him as she would speak to a child. [48]

Hellanicus, disturbed by chronological difficulties, had put Heracles in the generation previous to Theseus, [49] and so encouraged the belief that the tales of Theseus were merely Athenian adaptations of the legend of Heracles. Philochorus rejects this version altogether and makes Theseus a partner, not an imitator, of Heracles in his heroic efforts to conquer lawlessness. Although he was rescued from Aedoneus by Heracles and consecrated shrines to him in Athens as an expression of his gratitude, they go on the expedition against the Amazons together as equals. [50] But Theseus is not only a warrior on the Homeric model. Philochorus records that suppliants of all kinds took refuge in the Theseum at Athens, [51] a custom which establishes Theseus as a champion of the oppressed. He helped Adrastus in arranging a truce so as to recover the dead bodies after the expedition against Thebes; this is said to be the first truce ever arranged for such a purpose; [52] Theseus is therefore to be held responsible for this humane rule of ancient warfare.

Although Philochorus may reject versions of legends which are grotesque or discreditable, he shows his loyalty to the traditional style of the Atthidographers by finding aetia in plenty. He derived the name of the Pelasgians from their migratory habits, since in this respect they resembled birds and especially storks (πελαργοί); he said that the Pelasgians on Lemnos were called Sinties because they made plundering raids (σίνεσθαι) and derived the word τύραννος from Τύρρηνοι. [53] According to his account, when Cecrops wanted to find out how many subjects he had, he ordered each individual to de-

[47] Fg. 46.
[48] Fg. 37.
[49] F.166 (*FGrH* 1).
[50] Fg. 45, 49.
[51] Fg. 47.
[52] Fg. 51.
[53] Fg. 5–7. For the Sinties cf. Chap. 1, p. 13 above.

posit a stone in a certain place; in this manner he was able to derive the term for "people" (λαός) from λᾶας, "a stone"; [54] and his story provided an ancient precedent for the census of Attic citizens taken by Demetrius of Phalerum.   There are several other explanations in the same manner; [55] but there is little to suggest an interest in affairs outside of Attica; he mentions the Spartan veneration for Tyrtaeus only because the poet is supposed to be of Athenian origin. [56]

His treatment of Theseus and his attempt to derive Tyrtaeus from Athens reveal a special sort of Athenian patriotism, which is confirmed by a few fragments relating to later times.   Herodotus [57] tells how the Alcmaeonid family, after it had been exiled by the Peisistratids and defeated in battle at Leipsydrion, built a magnificent new temple of Apollo at Delphi, and in this way persuaded the oracle to take up the cause of democracy, so that it urged the Spartans, whenever they consulted Delphi about any project, to "set Athens free first."   This tale of bribery was felt as a slur not only on Delphi but also on the Alcmaeonids, who from time to time had difficulty in explaining certain incidents in their family history. [58] Philochorus, however, does his best to clear both parties of blame. "The story is," writes the scholiast on Pindar, "that the Delphic temple was burned down (by the Peisistratids, as they say), [59] and that the Alcmaeonids, exiled by the tyrants, promised to rebuild it; so they were given money, and, having collected an army, they attacked the Peisistratids; they were victorious and, besides many other thank-offerings, rebuilt the temple for the god which they had promised, as Philochorus relates." [60]   This version makes the Peisistratid reputation even blacker than usual; the Alcmaeonids are cleared of the charge of treason, since they do not invite the aid of Sparta, and all that Delphi does is to lend money in a just cause. Later on again, we find Philochorus defending the memory of the Alcmaeonids, when he makes the Corinthians responsible for the mutilation of the Hermae and acquits Alcibiades of all blame. [61]

[54] Fg. 12.   Cf. the etymology of ἄστυ in Fg. 4.

[55] Cf. the derivation of *Boedromia* from βοηθεῖν δρόμῳ in Fg. 33; he refuses to connect Athena's name *Sciras* with the sunshade (σκίρον) carried over her statue and prefers the derivation from a certain Scirus (Fg. 42).

[56] Fg. 55, 56.   Adrastus and his chariot at Harma in Boeotia (Fg. 50, 51) are relevant because Theseus intervened in the war of the Epigoni against Thebes.

[57] 5.62–63.

[58] I have discussed these occasions in *CPh* 31 (1936) 43–46.

[59] Hdt. 2.180 says that it was burned down accidentally (αὐτόματος κατεκάη).

[60] Fg. 70—Sch. Pi. *P.* 7.9.   Cf. Isoc. *Antid.* 232.

[61] Fg. 110.

His treatment of Nicias, on the other hand, seems to have been rather less favourable. He pointed out that the commander's superstitious fears were kept in check by Stilbides, the prophet who accompanied him to Sicily, but that this man unfortunately died before the fateful eclipse of the moon took place; this omen, Philochorus says, was really a favourable one, since enterprises like their flight "needed concealment." [62] These remarks should not be taken as an indication of his interest in anecdote or biographical detail, but rather as evidence of his own authority as a μάντις and an interpreter of omens. His authority in such matters is revealed by another anecdote: that, when the Persians occupied Attica, the dogs of the city set the citizens an example by attempting to swim over to Salamis. [63] This story is told to illustrate the validity of the oracles which advised the Athenians to desert their city. So also it appears that Procleides, "the ἐραστής of Hipparchus," is mentioned only because he was the first to dedicate a statue of three-headed Hermes. [64]

The priestly interest is equally evident in the remarks about Athenian topography, which are fairly plentiful. Philochorus evidently went to some trouble in tracing the origin of temples back to the time of Theseus, who founded a number of shrines in Athens, besides rededicating to Heracles the *Thesea* which the people had put up in his honour. [65] He also spoke of the *Araterion*, where Theseus pronounced solemn curses upon his political enemies. [66] He distinguished the different places near Athens called Colonus, evidently supplementing what Androtion had to say on the subject. [67] In later books, he described where the offerings of Meton the astronomer were set up, and the tripod of Aeschraeus above the theatre, with the inscription cut in the face of the rock; [68] and perhaps

[62] Fg. 112, 113.

[63] Fg. 84—Ael. *NA* 12.35. Cf. the tale about Xanthippus' dog in Plu. *Them.* 10.

[64] Fg. 69.

[65] Fg. 45. Cf. Plu. *Thes.* 17: μαρτυρεῖ δὲ τούτοις ἡρῷα Ναυσιθόου καὶ Φαίακος, εἰσαμένου Θησέως Φαληροῖ πρὸς τῷ τοῦ Σκίρου [ἱερῷ], καὶ τὴν ἑορτὴν τὰ Κυβερνήσιά φασιν ἐκείνοις τελεῖσθαι. Müller should have included this sentence in Fg. 41.

[66] Fg. 48—*EM* s.v. Ἀρητήσιον· τόπος Ἀθήνησιν οὕτω καλούμενος, ὅτι Θησεὺς μετὰ τὸ ὑποτρέψαι ἐκ τοῦ Ἅιδου, ἐκπεσὼν Ἀθηνῶν ἐκεῖσε τὰς κατὰ τῶν ἐχθρῶν ἀρὰς ἐποιήσατο. παρὰ τὰς ἀρὰς οὖν Ἀρητήσιον ὁ τόπος ἐκλήθη. οὕτω Φιλόχορος ἐν τῷ δευτέρῳ τῶν Ἀτθίδων. Cf. Plu. *Thes.* 35 (not cited by Müller): αὐτὸς δὲ Γαργηττοῖ κατὰ τῶν Ἀθηναίων ἀρὰς θέμενος, οὗ νῦν ἔστι τὸ καλούμενον Ἀρατήριον, εἰς Σκῦρον ἐξέπλευσεν.

[67] Fg. 73. Other sacred sites of which he spoke were the temple of Demeter Chloe (Stiehle, *Ph* 8 [1853] 638—Sch. Ar. *Lys.* 835) and the ἱεροὶ αὐλῶνες (Fg. 147).

[68] Fg. 99, 138.

how Speusippus dedicated some statues of the Graces on the Hill of the Muses, with an elegiac distich inscribed upon them (but this restoration is extremely uncertain).[69] His mention of the Lyceum as a gymnasium established in the time of Pericles shows his interest in buildings other than temples.[70] Phyle and Eetioneia [71] would naturally be mentioned in the course of his historical narrative, but he was not content to let the latter name pass without explaining that it took its origin from a certain Eetion.

These fragments are sufficient to show that, like the other Atthidographers, Philochorus interspersed remarks on Attic topography in the course of his narrative. There is no reason to suppose that he gave a continuous topographical description of Athens or Attica.[72]

Like his predecessors Philochorus also found occasion to comment on Athenian political and social institutions, though the references to them in the fragments are not so numerous as the references to religious customs. He said that armour was first made in the time of Cecrops, shields being made from the skins of wild animals.[73] He traces back the custom of mixing water with wine to Amphictyon, and explains that Dionysus was worshipped under the title of 'Ορθός, "the Upright," because this milder drink enabled people to stand upright, instead of stooping as they did under the influence of unmixed wine; he also mentions the custom of taking a sip of unmixed wine like a liqueur after meals, and finds a ritual significance in this practice.[74] He spoke of the "contest of manliness" (ἀγὼν εὐανδρίας) held at the Panathenaea, of the special cup called a *pentaploa* given to the victors at the festival of Athena Sciras, and of the rule against slaughtering a sheep unless it had been shorn once—a παλαιὸς νόμος of which Androtion also had spoken.[75]

Among Athenian political institutions, he spoke, naturally, of the Areopagus; how its powers in trying cases of homicide dated

[69] *Academicorum Philosoph. Ind. Herc.*, ed. Mekler, col. 6, 30–38.

[70] Fg. 96.

[71] Fg. 140, 115.

[72] The names of the demes (Fg. 71, 72, 74–78) would naturally be listed in an account of the reforms of Cleisthenes (cf. Böckh, *op. cit.* 411). Book III, the appropriate book, is cited in three of the fragments; ιγ' in Fg. 76 is most probably a mistake for γ' (cf. Chap. 4, p. 79, note 18, above).

[73] *POxy* 10.1241, col. 5, 6: Φιλόχορος δὲ καθόπλισιν γενέσθαι πρῶτον λέγει ἐπὶ Κέκροπος δόρυ καὶ δέρματος ἀγρίου περιβολήν. ὕστερον δ' ὅτ' ἤδη βόες ἐθύοντο βοέας τοὺς ἐν τῇ 'Αττικῇ ποιήσασθαι.

[74] Fg. 18, 19.

[75] Fg. 27, 43, 63, 64. Cf. Androt. Fg. 41.

from the trial of Ares; in what manner people were appointed to it
and how great a reputation it enjoyed; and how at one time it tried
people on charges of extravagant living beyond their apparent
means, as Phanodemus also had recorded.[76] He spoke also of the
ἀντιγραφεὺς τῆς διοικήσεως, of the "incapable persons" and the dole
paid to them, and of the Theoric Fund.[77] His explanation of the
procedure of ostracism is preserved *verbatim* and it is a much more
complete account than that given in Aristotle.[78] He also discussed
the organization of the γένη and the system of military organization
during the Peloponnesian War.[79] In later times, his remarks about
the symmories and about the νομοφύλακες and γυναικονόμοι established by Demetrius of Phalerum [80] would occur in the ordinary
course of historical narrative; they do not constitute proof that he
devoted special attention to the constitutional changes of the fourth
century.

It is fairly clear that in his interests and his choice of subjects for
digressions Philochorus conformed to the tradition established by
earlier writers of *Atthides*. He was ready to offer a new explanation
of the origin of some religious rite or festival, but ready also to
agree with an earlier authority. He found space in his narrative for
topographical and constitutional questions, just like the earlier
Atthidographers. He differed from them sometimes because he
had a more fastidious taste in *aetia* and a greater regard for the good
name of gods and heroes. It remains, therefore, to examine in

[76] Fg. 16, 17; 58, 59; 60. Cf. Phanod. Fg. 15.

[77] Fg. 61, 67–68, 85. Arist. *Resp. Ath.* 49, says the ἀδύνατοι received two obols a
day. According to Harp. *s.v.* ἀδύνατοι (Fg. 67) Philochorus said they received nine
drachmae a month. In Bekker's *Anecdota* 1.345, 15 (Fg. 68) the text reads: ἐλάμβανον
τῆς ἡμέρας, ὡς μὲν Λυσίας λέγει, ὀβολὸν ἕνα, ὡς δὲ Φιλόχορος πέντε, Ἀριστοτέλης δὲ δύο ἔφη.
Evidently the allowance varied at different times, but 5 obols a day is clearly a mistake;
9 drachmae a month would be equivalent to less than 2 obols a day.

[78] Fg. 79b—Phot. *Lex.*, ed. Porson, 675, 12ff: προχειροτονεῖ μὲν ὁ δῆμος πρὸ τῆς ὀγδόης
πρυτανείας εἰ δοκεῖ τὸ ὄστρακον εἰσφέρειν· ὅτε δὲ δοκεῖ (leg. δοκοίη), ἐφράσσετο σανίσιν ἡ
ἀγορὰ καὶ κατελείποντο εἴσοδοι δέκα, δι' ὧν εἰσιόντες κατὰ φυλὰς ἐτίθεσαν τὰ ὄστρακα,
στρέφοντες τὴν ἐπιγραφήν. ἐπεστάτουν δὲ οἵ τε ἐννέα ἄρχοντες καὶ ἡ βουλή. διαριθμη-
θέντων δὲ ὅτε (leg. ὅτου) πλεῖστα γένοιτο καὶ μὴ ἐλάττω ἑξακισχιλίων, τοῦτον ἔδει τὰ δίκαια
δόντα καὶ λαβόντα ὑπὲρ τῶν ἰδίων συναλλαγμάτων ἐν δέκα ἡμέραις μεταστῆναι τῆς πόλεως
ἔτη δέκα (ὕστερον δὲ ἐγένοντο πέντε), καρπούμενον τὰ ἑαυτοῦ μὴ ἐπιβαίνοντα ἐντὸς πέρα τοῦ
Εὐβοίας ἀκρωτηρίου. μόνος δὲ Ὑπέρβολος ἐκ τῶν ἀδόξων ἐξοστρακισθῆναι διὰ μοχθηρίαν
τρόπων, οὐ δι' ὑποψίαν τυραννίδος· μετὰ τοῦτον δὲ κατελύθη τὸ ἔθος, ἀρξάμενον νομοθετή-
σαντος Κλεισθένους, ὅτε τοὺς τυράννους κατέλυσεν, ὅπως συνεκβάλῃ καὶ τοὺς φίλους αὐτ.
. . . ." Cetera desunt. Cf. Arist. *Resp. Ath.* 43.5: ἐπὶ δὲ τῆς ἕκτης πρυτανείας πρὸς τοῖς
εἰρημένοις καὶ περὶ τῆς ὀστρακοφορίας ἐπιχειροτονίαν διδόασιν, εἰ δοκεῖ ποιεῖν ἢ μή.

[79] Fg. 91–93; 100, 101.

[80] Fg. 126, 129; 141a & b.

more systematic fashion the fragments of his historical narrative; thanks to the many references in Didymus and the scholiasts on Aristophanes, these are much more numerous than similar fragments of earlier *Atthides*.

There is one fragment which suggests that, like Hellanicus, he was interested in the chronology of early Attic history. He took from Hellanicus the figure of 1020 years as the interval between the time of Ogygus and the first Olympiad; and he is reported to have said that after Ogygus "the country now called Attica" remained without a king for 189 years till Cecrops; he denied that there was any such king as Actaeus and was content to derive the name of Attica from ἀκτή.[81] Very little else is preserved of his views on the chronology of early Attic history. He said that Cecrops reigned for fifty years and he placed the "floruit" of Homer forty years after the Ionian migration, 180 years after the Trojan War, when Archippus was archon at Athens.[82] These are indications that he wanted, so far as possible, to give an exact chronology of early times. But there is no way of telling whether he adopted Hellanicus' scheme of generations or his count of kings from Cecrops to Theseus.

Another fragment from his account of the period of the kings suggests a different kind of exactitude. He said that Cecrops, "wishing to know who were Athenians and how great their number was, gave orders that they were to take stones and deposit them in a certain place, in which way he discovered that they were twenty thousand in number."[83] This account of a census in very early times is quite clearly introduced to show that Demetrius of Phalerum had ancient precedent for his census of the Athenian people—which, by a very curious coincidence, showed the citizens to be just one thousand more in number than in the time of Cecrops![84]

We should be in a better position to pass judgment on the political purpose of this story if more fragments from his treatment of the sixth century were available. Unluckily, we have no useful

---

[81] Fg. 8 (Hellanic. F. 47a)—Eus. *PE* 10.10.7–8, quoting from Julius Africanus. Cf. Chap. 1, pp. 11–13 above.

[82] Fg. 10; 52–54a. For Philochorus as an authority on the date of Homer see also Eusebius *PE* 10.11.3. De Sanctis ('Ατθίς, ed. 2, 99–116), in his attempt to show that the list of rulers at Athens in Julius Africanus is derived ultimately from Philochorus, can find no real evidence except this mention of Archippus, "one of the archons appointed for life." It does not seem necessary to give a detailed account of his arguments here.

[83] Fg. 12. For the etymological point of the story see pp. 115–16 above.

[84] Ath. 6.272C.

information about his treatment of Solon and Cleisthenes. Suidas, in writing of the Seisachtheia, gives the orthodox definition that by it the poor people were able to "shake off their burden," and then adds that, according to Philochorus, "their burden was voted away by decree." [85] The point of this remark is not clear at all, so that we do not know what contribution Philochorus made to the discussion about Solon nor which side he took in the controversy between Androtion and Aristotle.[86] It would be most interesting to know what changes took place in the attitude towards Solon in the interval between Aristotle and Philochorus.

With the beginning of the fifth century chronological indications in the fragments become much more frequent. We find the names of Athenian archons mentioned more frequently, just as in Aristotle's *Constitution of Athens*. But the mere mention of archons' names cannot be taken as proof that Philochorus gave a continuous annalistic account of events from the time of the Persian Wars. There is conclusive evidence that he did give an annalistic account of the period of the Peloponnesian War; the scholia on Aristophanes, which will be quoted subsequently, leave no doubt on this point.[87] The difficulty, as with Hellanicus, is to decide at what point he started to use an annalistic system. The first certain indication for the *Atthis* of Hellanicus is for the year 407–6; [88] a passage from Philochorus, referring to 438 B.C., is introduced with the formula, "Philochorus says that the following took place in the archonship of Theodorus" (Φιλόχορος ἐπὶ Θεοδώρου ταῦτά φησιν).[89] It is extremely likely that his use of this system starts with the fourth book, which began some time in the middle of the fifth century. This will be a useful hypothesis to bear in mind, while the evidence of the fragments is examined in detail.

In the first place, it should be noted that the events associated with archons' names in the first part of the fifth century are not political events. Philochorus said that the worship of Hermes

[85] Presumably this is what he means, though the Greek is not quite clear: Fg. 57—Suid. *s.v.* Σεισάχθεια· χρεωκοπία δημοσίων καὶ ἰδιωτικῶν, ἣν εἰσηγήσατο Σόλων. εἴρηται δὲ παρ' ὅσον ἔθος ἦν Ἀθήνησι τοὺς ὀφείλοντας τῶν πενήτων σώματι ἐργάζεσθαι τοῖς χρήσταις· ἀποδόντας δὲ οἱονεὶ τὸ ἄχθος ἀποσείσασθαι· ὡς Φιλοχόρῳ δὲ δοκεῖ, ἀποψηφισθῆναι τὸ ἄχθος.

[86] Cf. section on Androtion, pp. 83–84 above.

[87] Cf. also Fg. 106—Sch. Luc. *Tim.* 30, where προσθείς is certainly a mistake for προθείς.

[88] F.171. Cf. Chap. 1, pp. 24–25 above.

[89] Fg. 97. See below p. 124.

Agoraios was started in the archonship of Cebris, who is assigned tentatively either to 486 or to one of the very first years of the century.[90]   A date of this kind belongs to temple records, with which Philochorus ought to be familiar; it is no indication that political events were assigned to the year of Cebris.   His mention of Lacrateides (also tentatively placed at the opening of the fifth century) is no more helpful; [91] his year is connected with a great snowfall and an exceptional frost, for which and for no other reason his name would be remembered, just as many Americans will continue to remember 1888 as the year of the great blizzard and 1938 as the year of the hurricane.   Two dates from the 'Αρχόντων ἀναγραφή of Demetrius of Phalerum are in the same category: Demetrius mentioned Thales and Anaxagoras as beginning their professional careers in the archonships of Damasias (582–1) and Callias (480–79) respectively.[92]   These dates reveal an effort to establish the chronology of prominent literary men and scholars, in the manner of Eratosthenes and Apollodorus; but they cannot be taken as evidence for the previous existence, whether in an *Atthis* or elsewhere, of an annalistic record of political events.

No other chronological references are available for the first half of the century.   The fragments from Book IV give more information.   The scholiast on Aristophanes refers to Philochorus for an account of the two Sacred Wars; the names are evidently corrupted in some manuscripts, but, if the text is emended so as to harmonize with the account of Thucydides,[93] Philochorus says that the Spartan expedition to wrest the control of Delphi from the Phocians was followed "two years later" (τρίτῳ ἔτει) by the Athenian expedition which restored it to its former masters.   Thucydides, in his usual manner, says that the Athenian expedition took place "later on, after an interval" (αὖθις ὕστερον).[94]   But, as Beloch has shown,[95] there is no need to suppose any conflict between the two accounts or to reject the more accurate indication of date given by Philochorus.   This passage is not yet evidence that our author possessed

[90] Fg. 82.   Cf. above p. 112, note 36.
[91] Fg. 83—Sch. Ar. *Ach.* 220.   The Lacrateides in the text of the play may or may not have something to do with this old archon.
[92] F.1, 2 (*FGrH* 2 B, no. 228).
[93] Fg. 88— Sch. Ar. *Av.* 557.   For the different readings and emendations see J. W. White's edition of the scholia.   The account of Thucydides is in 1.112.5.
[94] Cf. Chap. 2, pp. 43–44 above.
[95] *Griech. Gesch.* 2.2.213.

special information about political dates; but it is a hint (though not certain proof) that he has started to use an annalistic system.

His remarks about the recovery of Euboea by Pericles in 445, as reported by the scholiast, do not seem to add anything to the account of Thucydides.[96] Another event which Philochorus assigned to the year 445–4, "when Lysimachides was archon," is the gift sent to the Athenians from Libya by Psammetichus of 30,000 medimni of grain, which was shared out among 14,240 citizens, after 4760 persons were shown to be claiming citizenship falsely.[97] No earlier authority is known for this incident, and Plutarch, though his figures are not quite the same,[98] probably learnt of it from him. The mention of this incident by Philochorus shows another attempt on his part to link up the history of his own times with the past, though presumably he has better authority for his statement here than for his account of the census taken by Cecrops. Not only is this gift of Psammetichus a precedent for the gifts of grain sent to Athens by Ptolemy I and others in the early part of the third century,[99] but the checking of citizenship claims corresponds to another recurring event in the lifetime of Philochorus: the revision of the citizen roll under the various anti-democratic régimes. These considerations naturally make us regard his account with some suspicion; but the possibility still remains that he had seen the text of the decree in the archonship of Lysimachides which ordered the distribution of the grain. In any case, this is the first exact date given by Philochorus for an event of political importance at Athens which is not recorded by any earlier authority known to us.

His account of the ostracism of Thucydides, son of Melesias,[100] is not clearly reported, since one commentator has confused exile and ostracism and another has not distinguished this Thucydides from the historian. Again, in recording his narrative of the dis-

---

[96] Fg. 89—Sch. Ar. Nu. 213. Dindorf prints as follows, borrowing the phraseology of Th. 1.114: Περικλέους δὲ στρατηγοῦντος καταστρέψασθαι αὐτοὺς πᾶσαν (sc. Εὔβοιαν) φησὶ Φιλόχορος· καὶ τὴν μὲν ἄλλην ἐπὶ ὁμολογίᾳ κατασταθῆναι, Ἑστιαιέων δὲ ἀποικισθέντων αὐτοὺς τὴν χώραν ἔχειν. For the date cf. Th. 2.2: τέσσαρα μὲν γὰρ καὶ δέκα ἔτη ἐνέμειναν αἱ τριακοντούτεις σπονδαὶ αἳ ἐγένοντο μετ' Εὐβοίας ἅλωσιν. τῷ δὲ πέμπτῳ καὶ δεκάτῳ ἔτει, ἐπὶ Χρυσίδος ἐν Ἄργει τότε πεντήκοντα δυοῖν δέοντα ἔτη ἱερωμένης καὶ Αἰνησίου ἐφόρου ἐν Σπάρτῃ καὶ Πυθοδώρου ἔτι τέσσαρας μῆνας ἄρχοντος Ἀθηναίοις κ.τ.λ.

[97] Fg. 90—Sch. Ar. V. 718. For the difficulties of the figures see A. W. Gomme, The Population of Athens, 16–17.

[98] Per. 37. He gives 14,040 instead of 14,240, and 40,000 instead of 30,000 medimni.

[99] Ferguson, Hellenistic Athens 141, 147.

[100] Fg. 95—Sch. Ar. V. 947.

grace and death of Pheidias, the scholiast has given wrong names to the archons, but probably the passage may be correctly translated as follows:

> Philochorus, under the heading of the archon Theodorus (438–7), writes as follows: "And the golden statue of Athena was set up in the great temple, containing gold to the weight of forty-four talents; [101] the overseer being Pericles and the artist Pheidias. And Pheidias, the artist, being under suspicion of misappropriating some of the ivory used for the scales on the serpents, was brought to trial and found guilty. And it is said that, having fled to Elis, he there undertook the contract for the statue of Zeus at Olympia. And after completing it he was put to death by the Eleans on the charge of misappropriating material, in the archonship of Pythodorus (432–1)." [102]

This translation contains a few explanations that are offered by another scholion in a paraphrase of the passage; but in other respects it is a fairly literal version, which preserves the simple chronicle style of the original; there are no subordinate clauses and the connectives are restricted to καί and δέ. It should be noted that the guilt or innocence of Pheidias is left for the reader to decide.

The scholiast on Aristophanes cites Philochorus again for the expeditions to Sicily during the Archidamian War; [103] and he quotes a fine example of annalistic description for the year of Euthynus (426–5): "The Spartans sent delegates to the Athenians to discuss a peace settlement, after making a truce with the men at Pylos and handing over their ships, which numbered sixty. When Cleon opposed the settlement, it is said that the assembly was split into two factions. Finally the president put the question to the vote; and those who wished to fight on carried the day." [104]

The scholiast on Aristophanes continues to cite him for the dates of various events through the period of the war. He put the revolt

---

101 Forty talents is the weight given by Th. 2.13.5.

102 Fg. 97—Sch. Ar. *Pax* 605. The Greek is worth quoting in full as an example of Philochorus' style: Φιλόχορος ἐπὶ Θεοδώρου ἄρχοντος ταῦτά φησιν· " Καὶ τὸ ἄγαλμα τὸ χρυσοῦν τῆς Ἀθηνᾶς ἐστάθη εἰς τὸν νεὼν τὸν μέγαν, ἔχον χρυσίου σταθμὸν ταλάντων μδ´, Περικλέους ἐπιστατοῦντος, Φειδίου δὲ ποιήσαντος. καὶ Φειδίας ὁ ποιήσας, δόξας παραλογίζεσθαι τὸν ἐλέφαντα τὸν εἰς τὰς φολίδας, ἐκρίθη. καὶ φυγὼν εἰς Ἦλιν ἐργολαβῆσαι τὸ ἄγαλμα τοῦ Διὸς τοῦ ἐν Ὀλυμπίᾳ λέγεται, τοῦτο δὲ ἐξεργασάμενος ἀποθανεῖν ὑπὸ Ἠλείων ἐπὶ Πυθοδώρου, ὅς ἐστιν ἀπὸ τούτου ἕβδομος." The emendations Θεοδώρου (archon 438–7) instead of Πυθοδώρου, and Πυθοδώρου (432–1) for Σκυθοδώρου are due to Palmer. See Dindorf's edition of the scholia. There is similar confusion over archons' names in Sch. Ar. *Pax* 990; this frag. is not in *FHG;* see Strenge, *Quaestiones Philochoreae* 65.

103 Fg. 104—Sch. Ar. *V.* 240.

104 Fg. 105—Sch. Ar. *Pax* 665.

of Scione, at Brasidas' instigation, in the year of Isarchus (424–5), the Peace of Nicias in the year of Alcaeus (422–1), and the condemnation of the mutilators of the Hermae in the year of Chabrias (415–4).[105] He recorded that the Athenians first began to use Athena's silver under Callias (412–1)—a detail that every Exegetes would know; that Cleophon rejected the Spartan offer of peace under Theopompus (411–0); and that a gold coinage from the golden Victories was issued under Antigenes (407–6).[106] For the year 410–9 he recorded: "Under Glaucippus also the Boule for the first time was seated in alphabetical order, and beginning from that time its members still swear to keep the seat assigned to them according to the letter of their name." [107] Thus it appears that in his account of the rigours of the Peloponnesian War he did not neglect to record slight details of official procedure, especially if he thought such changes (like this one) showed the origin of customs of his own time. Finally, we discover that he gave the names of the unlucky Athenian generals at Arginusae; that, like Aristotle, he reported the appointment of thirty συγγραφεῖς in 411 to draw up constitutional reforms (as opposed to Thucydides who mentioned only the ten πρόβουλοι first appointed); and his account of the death of Critias is recorded, though in hopelessly corrupt form.[108]

These fragments from his chronological account of events in the fifth century have been set forth in some detail in order to show how little evidence there is to justify the belief that Philochorus had literary sources at his disposal which were not available to Thucydides or Hellanicus. He knows the dates of several events of religious significance, which are not recorded elsewhere; it would not be surprising if knowledge of many such dates, even for much earlier events, was claimed, whether correctly or not, by priestly colleges. On the other hand, Psammetichus' gift of grain to Athens had a special significance for an Athenian writer in the first half of the third century. The date of this gift is not recorded elsewhere; but it is the only important date of political significance in the fifth century, previous to the Peloponnesian War, that is recorded for

---

[105] Fg. 107, 108, 111—Sch. Ar. V. 210, Pax 468, Av. 766.

[106] Fg. 116—Sch. Ar. Lys. 173; Fg. 117, 118—Sch. Eur. Or. 371, 772; Fg. 120—Sch. Ar. Ra. 720. In this last instance, if Bentley's conjecture is correct, he is following Hellanicus (see Chap. 1, p. 14, note 50).

[107] Fg. 119—Sch. Ar. Pl. 972.

[108] Fg. 121; 122—Harp. s.v. συγγραφεῖς (cf. Arist. Resp. Ath. 29, 2; Th. 8.67); Fg. 123.

us by Philochorus alone. No attempt will be made here to solve the problem of Aristotle's chronological source for the fifth century; but it should be pointed out that those who think he obtained his information ultimately from the same source as Philochorus, from an *Ur-Atthis* by some unknown author, have no real evidence apart from this fragment. Admittedly, their theory cannot be disproved; but it should be recognized on how uncertain a basis it rests, since Philochorus may have obtained even this date from an independent, perhaps an epigraphical source.

There is no necessity to discuss the fragments relating to the political history of the fourth century from the same point of view. We know that Philochorus could consult detailed accounts of the first half of the century—those of Ephorus, Theopompus, and the Oxyrhynchus historian among others. But the question of his sources for the history of the fourth century does not concern us here; our discussion will be restricted to the style and method of his narrative.

The passages quoted by the scholiast on Aristophanes show his use of a concise, annalistic style in recording events of the second half of the fifth century and especially in his account of the Peloponnesian War. His use of similar methods for the fourth century is proved by quotations in Didymus and Dionysius of Halicarnassus. The passage which apparently described the movements of Conon leading up to the battle of Cnidus is sadly mutilated in the text of Didymus and cannot be used as evidence; [109] but there are several fragments referring to the events which followed upon this battle. Didymus refers briefly to his account of the rebuilding of the long

[109] Did. *In D.* Col. 7, 36–51. A more complete restoration than that of Diels and Schubart is offered by De Gubernatis, *Aegyptus* 2.23–32: Κόνων μὲν ἀπὸ Κύπρου μετὰ πασῶν τῶν νεῶν ἀπῆρε τῷ δὲ τῆς Φρυγίας σατράπῃ βουλόμενος εὐθὺς συμμεῖξαι καὶ εἰς τὸ ναυτικὸν χρήματα λαβεῖν. ἐπ' Εὐβουλίδου δὲ ἔπλευσεν ἐκ Ῥόδου μετὰ ὀγδοήκοντα μὲν τριήρων ἀπὸ Φοινίκης, δέκα δ' ἀπὸ Κιλικίας. . . . 3 lines with only a few letters on the papyrus follow. Then: τελευταῖον δὲ τὰς ναῦς συνέλεξε πρὸς Λώρυμα τῆς Χερσονήσου καὶ ἐντεῦθεν ἐκπλεύσας καὶ ἐπιπεσὼν τῷ στόλῳ τῶν Λακεδαιμονίων. . . . καὶ ναυμαχίας γενομένης ἐνίκησε καὶ πεντήκοντα τριήρεις αἰχμαλώτους ἐποίησε καὶ Πείσανδρος ἐτελεύτησεν. De Gubernatis remarks on the "excessive haste" of Philochorus, so that not every stage of the narrative is made clear to the reader. If his restoration is correct, Philochorus told the story of Conon's movements from 397–94 without interruption or reference to events elsewhere—adopting the method which Ephorus used for Cimon's campaigns (see Chap. 5 p. 96) and which at times confused Diodorus. In that case it should be said that he combined the κατὰ γένος method with the strictly annalistic; but one is unwilling to conclude so much from a restored passage.

walls at Athens;[110] but his account of the events of 392–1 is more illuminating.

Didymus discusses at some length what Demosthenes meant when, in the *Fourth Philippic*, he spoke of the help given by the Persian King to the Athenians in earlier days.[111] He first mentions the view that Demosthenes was referring to the terms of peace offered by the King through Antalcidas in 392–1. "Philochorus," he says, "gives a description in these very words, under the heading of the archon Philocles (392–1): 'And the King sent down the peace negotiated by Antialcidas (*sic*), which the Athenians refused to accept, because it was written in it that the Greeks living in Asia should all be reckoned as in the estate of the King.'"[112] This account differs in one important detail from Xenophon's version.[113] According to Xenophon, the Athenians refuse to accept the terms which are proposed by Antalcidas at a conference with Tiribazus; and when Tiribazus, after arresting Conon, goes to make his report to the King about the proposals of Antalcidas and to ask for instructions, the King, instead of "sending down a peace," "sends down Strouthas," who is to prosecute the war at sea against Sparta with vigour. Besides the failure of these peace negotiations with Persia in 392–1, Philochorus also described the abortive negotiations for peace with Sparta in the same year when "ambassadors came from Sparta and departed without accomplishing anything, since Andocides could not convince the Spartans."[114]

Didymus rejects the view that Demosthenes is referring to the peace offer of 392–1 and thinks that, in speaking of the support

---

[110] Col. 7, 51–54: ἀπὸ δὲ ταύτης τῆς ναυμαχίας ὁ Κόνων καὶ τὰ [μακρὰ τ]είχη τοῖς Ἀθηναίοι[ς] ἀνέστησε[ν ἀκόν]των Λακεδαιμονίων, καθάπερ πάλιν ὁ αὐτὸς συγγραφεὺς ἱστο[ρ]εῖ.

[111] Dem. 10.34.

[112] Diels and Schubart print as follows (col. 7, 17–30): Φιλό]χορος ἀφη[γεῖ]τ[α]ι αὐτοῖς ὀνό[μ]ασι, πρ[οθεὶ]ς ἄρχοντα Φιλο[κλέ]α Ἀναφλύ[σ]τιον· " Καὶ τὴν εἰρήνην τὴν ἐπ' Ἀντ[ια]λκίδου κατέπ[ε]μψεν ὁ βα[σ]ιλεύς, ἣν Ἀθηναῖοι ο[ὐκ] ἐδέξαντο, δ[ι]ότι ἐγέγ[ρ]απτο ἐν αὐτῇ τοὺ[ς τὴν Ἀσ]ίαν οἰκοῦντ[ας] Ἕλληνας ἐν βασιλέως οἴκ[ῳ π]άντας εἶναι [σ]υννενεμη-μένους. ἀλλὰ καὶ τοὺ[ς πρέσ]βεις τοὺς ἐν Λακεδαίμονι συγχωρήσα[ντας] ἐφυγάδευσα[ν] Καλλιστράτου γράψαντος [καὶ οὐ]χ ὑπομείναντας τὴν κρίσιν Ἐπικράτην Κ[η]φισιέα, Ἀνδοκ[ί]δην Κυδαθηναιέα, Κρατῖνον .[.].ιστιον, Εὐβο[υ]λίδην Ἐλευσίνιον." οὐκοῦν ὅτι μὲν οὐκ εἰκός ἐστι [τ]ὸν [Δ]ημοσθένη ταύτης αὐτοὺς ὑπομιμνή[σκ]ειν τῆς ε[ἰ]ρήνης ἑόραται. It seems likely, however, that the quotation from Philochorus really ends at συννενεμη-μένους, and that the next sentence is part of Didymus' argument; he uses ἀλλὰ καί in argument in col. 7, 15 and probably also in 5, 65.

[113] *HG* 4.8.12–17.

[114] Argumentum in Andoc. *De Pace*—*FHG* 4.646. For the exile of Andocides, following on his diplomatic mission to Sparta, cf. *Vitae X Oratorum* 835A (*Biogr. Gr.*, ed. Westermann, p. 239).

given by the King to the Athenians, he means his help of Conon before the battle of Cnidus.[115]   He then adds that possibly Demosthenes is thinking of another offer of peace made by the King "which they accepted gladly; and again Philochorus has spoken of this, how, although it was similar to the proposals made by the Spartan Antalcidas, they accepted it since they had spent all their money keeping up armies of mercenaries and had for a long time been in a state of exhaustion because of the war; it was on this occasion that they set up the altar of Peace." [116]   Philochorus, after his manner, does not forget to mention the inauguration of this new religious cult, which, according to one account recorded in Plutarch,[117] was first introduced when the Peace of Callias was made.   On a point of this kind one is inclined to trust Philochorus rather than another authority.   Didymus goes on to say that "one might mention many other services which the King rendered to the city, as, for example, the peace negotiated by Callias, son of Hipponicus." But one cannot take these words as an indication that Philochorus spoke of the Peace of Callias.

The four fragments assigned to Book V by Müller are not direct quotations and add nothing to our knowledge of Philochorus' style. They show simply that he mentioned the alliances that led up to the Corinthian War,[118] the first establishment of the symmories in 377, and Philip's early activities in Thrace.[119]   Another fragment shows that he mentioned the sending of Athenian cleruchs to Samos in the archonship of Aristodemus (352–1);[120]   most modern historians prefer to date this event in 361–0, following the scholiast on Aeschines.[121]

Both Philochorus and Androtion give very similar reports of an event of the year 350–49, the settlement by the Athenians of the

---

[115] Col. 7, 30–34.   He then quotes Philochorus' account of events leading up to the battle (see note 109 above).

[116] Col. 7, 62–71.

[117] *Cim.* 13.   Isocrates, 15.109–10, says that yearly sacrifices to Peace first took place after the peace with Sparta which followed Timotheus' victory at the battle of Leucas, and Nep., *Timoth.* 2.2, interprets this to mean that altars to Peace were first set up at that time.

[118] Fg. 125—Sch. Ar. *Ec.* 193.   The text reads: περὶ τοῦ συμμαχικοῦ Φιλόχορος ἱστορεῖ, ὅτι πρὸ δύο ἐτῶν ἐγένετο συμμαχία Λακεδαιμονίων καὶ Βοιωτῶν.   As Müller points out, Λακεδαιμονίων is clearly a mistake, probably for 'Αθηναίων.

[119] Fg. 126, 127, 128.

[120] Fg. 131—D.H. *Dein.* 13.

[121] Sch. Aeschin. *In Timarch.* 53: εἰς Σάμον κληρούχους ἔπεμψαν 'Αθηναῖοι ἐπ' ἄρχοντος Νικοφήμου.   Cf. Beloch, *Griech. Gesch.* 3.1.194.

boundaries of the ἱερὰ ὀργάς in the Megarid, a tract of land sacred to Demeter and Persephone. "This division," says Didymus,[122] "took place in the archonship of Apollodorus, as Philochorus relates. He writes as follows: 'The Athenians had quarrelled with the Megarians over the boundaries of the sacred Orgas and entered Megarian territory with an army under Ephialtes, who was strategus at the time, and established the boundaries of the sacred Orgas. And at the consent of the Megarians those who established the boundaries were Lacrateides the hierophant and the *dadouchos* Hierocleides. They declared the land at the edges of the Orgas sacred, since the oracle had declared that "it will be better and more fortunate for you if you leave the fields idle and do not till them." And they marked off the boundaries all round with stelae according to the decree of Philocrates.'" Since this is a religious matter, one looks for an authoritative account in Philochorus. It is disappointing to find that he has apparently done no more than reproduce Androtion's description, using mostly the same words and altering the phrasing only slightly.[123] He seems to have followed Androtion in a similar manner in his account of the διαψηφίσεις of 346–5.[124]

Dionysius of Halicarnassus quotes several passages from Book VI describing events in the struggle with Philip, which offer good examples of the style of the *Atthis*. Dionysius quotes the following passage for "the beginning of the war about Olynthus" in the archon year 349–8: "Callimachus from Pergase. In this man's archonship, when the Olynthians were attacked by Philip and sent delegates to Athens, the Athenians made an alliance with them and sent them help, to the extent of two thousand peltasts, the thirty triremes under Chares, and eight others which they equipped and manned." [125] Immediately afterwards Dionysius quotes another sentence from his description of the same year, which is introduced

---

[122] Col. 13, 44–58. He refers to this event for the purpose of dating Oration 13. The reading of the papyrus Φιλόδωρος is quite clearly a mistake for Φιλόχορος.

[123] Androtion's account is in col. 14, 35–49. Since both authors mention Philocrates as having prepared the *psephisma*, it is possible that Androtion, if not Philochorus also, had actually seen the stone on which it was recorded. The repetitious language in the text certainly recalls the style of an official document.

[124] Fg. 133—Harp. s.v. διαψήφισις· . . . ἐντελέστατα δὲ διείλεκται περὶ τῶν διαψηφίσεων, ὡς γεγόνασιν ἐπὶ Ἀρχίου ἄρχοντος, Ἀνδροτίων ἐν τῇ Ἀτθίδι καὶ Φιλόχορος ἐν ἕκτῳ τῆς Ἀτθίδος.

[125] Amm. 9—Fg. 132. The reading of the Teubner text has been adopted: τριήρεις δὲ τριάκοντα τὰς μετὰ Χάρητος καὶ ἃς συνεπλήρωσαν ὀκτώ.

by the familiar Thucydidean phrase "about the same time" (περὶ τὸν αὐτὸν χρόνον).[126]   Another passage follows, in similar Thucydidean narrative style, describing the Athenian response to the final appeal of the Olynthians in the spring of 348: "And again when the Olynthians sent delegates to Athens and begged that the Athenians would not suffer them to be defeated but would send them help in addition to the forces already present, help consisting not of mercenaries but of Athenians, the people sent them seventeen more triremes and twenty thousand citizen hoplites and three hundred horsemen in horse-transport ships, with Chares as commander of the entire force."   No doubt these passages were quoted by Didymus in his commentary on the Olynthiac orations, which is not preserved on the papyrus.

Didymus refers both to Androtion and Philochorus in his account of how the Athenians rejected the peace proposed by the Persian king in 344–3; he quotes only this concise and brief description of Philochorus: "In this year, when the King sent ambassadors to Athens and asked that they should continue towards him the friendship they had shown his father, they replied to the ambassadors at Athens that the King would continue to have their friendship if he did not attack Greek cities."[127]   Didymus goes on to point out the discourteous nature of this reply, but does not suggest that any such criticism of it was offered by the historian. Again, he quotes Philochorus' summary of events in 342–1, when the Athenians recovered the allegiance of Euboea and reestablished the democracy in Oreus: "And the Athenians made an alliance with the Chalcidians, and set free the people of Oreus with the aid of the Chalcidians in the month of Scirophorion, with Cephisophon as general in command; and the tyrant Philistides died."   A passage in similar style from his account of the next year follows: "In this year the Athenians crossed over to Eretria with Phocion as general in command, and with the object of restoring the democracy they besieged Cleitarchus, who formerly had been a rival political leader of Plutarchus and worked against him and after his exile established himself as tyrant; but now the Athenians forced him to submit after a siege and restored the city to the people."[128]

Dionysius of Halicarnassus quotes extensively from his account

---

[126] This phrase is used also in Fg. 146.
[127] Col. 8, 14–23. See also Chap. 4, p. 78 above.
[128] Col. 1, 13–25.

of the year 340–39: "The reasons on account of which they (the Athenians and Philip) went to war, each side claiming to have been unfairly treated, and the date when they broke the peace are accurately described by Philochorus in his sixth book. I will quote the most important passages from his account: 'Theophrastus of Halae. In this man's archonship Philip first sailed up to Perinthus and attacked it; and meeting with no success there, he laid siege to Byzantium and brought up siege engines.' Then, after describing the protests of Philip in his letter to the Athenians, he goes on as follows: 'The people, after hearing the letter, with Demosthenes urging them to go to war and proposing the necessary decrees, voted to pull down the stele which had been set up to record the peace and alliance with Philip, to man ships, and to take the other steps preparatory to fighting.'" [129]

At this point one must interrupt the narrative of Dionysius in order to quote another passage from Didymus.[130] Didymus, after mentioning Philip's attacks on Perinthus and Byzantium and his reasons for wanting to detach these cities from their friendship with Athens, describes the "entirely unjustifiable action" of Philip in seizing the merchant ships off Mount Hieron, "230 according to Philochorus, but 180 in the account of Theopompus,[131] from which he gathered up 700 talents; these things he has just recently done in the archonship of Theophrastus (340–39), successor to Nicomachus, as among other writers Philochorus records as follows: 'And Chares set out to join the meeting of the King's generals, leaving his ships off Mount Hieron, so that they could act as a convoy to the merchant vessels coming from the Pontus. Philip, when he found that Chares was gone, first of all attempted to send his ships so as to drive the vessels to shore; but not being successful in this way, he brought soldiers over to the opposite shore of the bay and so captured the vessels.[132] These were altogether not less than 230. And sorting out those which belonged to the enemy, he broke them

---

[129] *Amm.* 11—Fg. 135. This last passage is also quoted by Didymus, col. 1, 70–2, 2. Sylburg's text of Dionysius, which Müller follows, is incomplete in introducing this second quotation.

[130] Col. 10, 34—11, 5.

[131] This is not the only occasion when Philochorus gives different figures from Theopompus. Cf. Fg. 103 for their disagreement about the length of Perdiccas' reign.

[132] This sentence is not entirely satisfactory and the text may be at fault: τὸ μὲν [π]ρῶτον ἐπειρᾶτο πέμψαι τὰς ναῦς τὰ [π]λοῖα καταγαγεῖν· οὐ δυνάμενος δὲ βιάσα[σ]θαι στρατ[ι]ώτας διεβίβασεν εἰς τὸ πέραν ἐ[φ'] Ἱερὸν καὶ τῶν πλοίων ἐκυρίευσεν.

up and used the timber for his siege works; he also obtained a large quantity of grain, hides, and money.'"

We can return now to Dionysius, who says that Philochorus describes the events of the next year after the repudiation of the peace: "And I will give the most important passages from his account: 'Lysimachides of Acharnae. In his archonship the Athenians postponed their work on the docks and the storehouse for equipment because of the war with Philip. And they passed a decree, proposed by Demosthenes, that all available money be devoted to war purposes.' Then, when Philip had captured Elatea and Cytinium and sent ambassadors to Thebes representing the Thessalians, Aenianians, Aetolians, Dolopes, and Phthiotes, and the Athenians at the same time sent Demosthenes and the others, they voted to make an alliance with these people." [133]

This last sentence is a summary rather than a quotation from Philochorus, though it begins by quoting the historian's own words. The whole sentence was quoted by Didymus, but, since the text fails before the end and the portion which is preserved seems to be very clumsily constructed, it is not worth while to attempt a translation.[134]

A note in the *Lives of the Ten Orators* [135] records that Philochorus described the death of Demosthenes by poison, but there is nowhere in the fragments relating to the struggle with Philip any suggestion that he passed judgment on the merits of his policy or at any point went beyond a bare narrative of the events. A few fragments of antiquarian interest, which were discussed earlier in this chapter, show that he allowed himself digressions from his annalistic record. But if his account had contained any discussion of a critical nature, Didymus would certainly have found occasion to quote it   At one point, in commenting on the opening sentence of Oration 11, Didymus remarks that Philip's reason for attacking Perinthus and Byzantium was "to intercept the Athenian grain-route from the Black Sea, and to ensure that they should not have cities on the

---

[133] The translation is not quite certain here.

[134] Diels and Schubart print as follows (col. 11, 37–51): καὶ Φιλόχ[ορος δ' [ὅτι] Λ[οκ]ροῖς Φ[ί]λιππος αὐτὴν ἐκέ[λευσε] π[ρὸ]ς Θη[η]βαίων ἀποδοθῆναι διὰ τῆ[ς] ἔ[κ]τ[ης] φησὶ τὸν τρόπον τοῦτον· " Φιλ[ίππου] δὲ [καταλα]βόντος Ἐλάτειαν καὶ Κυτίν[ιον] καὶ πρέσβ[ε]ις πέμψαντος εἰς Θήβας Θε[ττα]λῶν Αἰν[ι]άνων Αἰτώλων Δολόπων Φθιωτῶν καὶ ἀξιοῦντος Νίκαιαν Λοκροῖς παραδιδόναι παρὰ τὸ δόγμα τὸ τῶν ἀμφικτυόνων, ἣν ὑπὸ Φιλίππου φρουρουμένην, ὅτ' ἐκεῖνος ἐν Σκύθαις ἦν, ἐκβαλόντες [τ]οὺς φρουροὺς αὐτοὶ κατεῖχον οἱ Θηβαῖοι, τούτοις μὲν ἀπεκρίναντο πρεσβείαν ὑπ[ὲρ] ἁπάντων πρὸς Φίλιππον διαλεξομένην < ... >."

[135] 847A—Fg. 139. Cf. 846B (*Biogr. Gr.*, ed. Westermann, p. 287).

coast which they could use as naval bases against him." [136] There
is no suggestion that he has borrowed this argument from Philochorus, and the very constructions which he uses are more com-
plicated than anything in the fragments which he quotes.

Everything goes to show, therefore, that Philochorus' account of
this period was a bald annalistic record, in extreme contrast with
the highly rhetorical treatment of Theopompus. At the same time,
it was evidently an accurate account and convenient for purposes of
reference, since Didymus prefers to quote from it rather than from
other authors with whom he seems to be familiar.

Whereas Book VI covered a period of at least twenty, possibly
as much as forty years, the later books are on quite a different scale.
Unluckily, the fragments here are much scantier and there is really
no way of telling to what degree Philochorus altered his method and
style. The lexicographers, apparently, thought he gave the best
available account of the magistracies instituted by Demetrius of
Phalerum, but the sentence which Athenaeus quotes about the
γυναικονόμοι does not suggest any great abundance of detail.[137]

Of the two passages quoted by Dionysius of Halicarnassus from
Books VIII and IX, the latter has already been quoted and discussed.[138] It shows how in the year 292 Philochorus interpreted
certain omens as indicating the return of exiles, among whom was
the orator Deinarchus. Nothing more need be said about this
passage here, except that it shows how the annalistic system was
continued and the sentences were still constructed on the same plan,
with simple participial phrases and clauses connected with καί or δέ.
The earlier passage, from Book VIII, is quoted by Dionysius, after
he has completed his account of the life of Deinarchus, in order to
show the authority on which his narrative rests. "Such," he
writes, "was the life of the orator; and each detail is established by
the histories of Philochorus and what Deinarchus has recorded
about himself in the speech against Proxenus. . . . Philochorus in
his *Attic History* describes the exile of the anti-democratic party
(i.e. the adherents of Demetrius of Phalerum) and their subsequent
return as follows: 'As soon as Anaxicrates entered on his archonship the city of Megara was captured. And King Demetrius,

[136] Col. 10, 40–45.
[137] 6.245C—Fg. 143: οἱ γυναικονόμοι, φησί, μετὰ τῶν Ἀρεοπαγιτῶν ἐσκόπουν τὰς ἐν
ταῖς οἰκίαις συνόδους, ἔν τε τοῖς γάμοις καὶ ταῖς ἄλλαις θυσίαις.
Cf. Fg. 141, 142.
[138] See above p. 107.

returning from Megara, made his preparations against Munychia and after tearing down the walls restored the city to the people. Subsequently a large number of the citizens were impeached, including Demetrius of Phalerum; those of them who did not wait to stand trial they condemned to death by vote, but those who submitted to trial they acquitted.'" [139]

Not only is this passage in his usual general style, but there are certain phrases in it which he uses elsewhere. His statement that they condemned some people "without their standing trial" (τοὺς μὲν οὐχ ὑπομείναντας τὴν κρίσιν ἐνεθανάτωσαν τῇ ψήφῳ) recalls his description of the banishment of Andocides and the other envoys to the peace conference at Sparta: ἀλλὰ καὶ τοὺς πρέσβεις τοὺς ἐν Λακεδαίμονι συγχωρήσαντας ἐφυγάδευσαν Καλλιστράτου γράψαντος καὶ οὐχ ὑπομείναντας τὴν κρίσιν. . .[140] Again, his remark about Demetrius Poliorcetes that he "restored the city to the people" (ἀπέδωκε τῷ δήμῳ) occurs also in his description of the expulsion of Cleitarchus from Eretria: τότε δὲ ἐκπολιορκήσαντες αὐτὸν Ἀθηναῖοι τῷ δήμῳ τὴν πόλιν ἀπέδωκαν.[141] There is, of course, nothing unusual about this phrase in itself. But it is interesting to note that Philochorus is content to describe the changes that took place after the flight of Demetrius of Phalerum simply by saying "they gave back the city to the people"—a conventional phrase, more suitable to a summary than to a detailed narrative. The use of such conventional and bald language to describe a political upheaval is evidence of the lack of distinction in his style.

It is also clear that he passed no judgment on the merits of Deinarchus or the party to which he belonged, since Dionysius could scarcely have failed to quote any such verdict if he had found it. The only occasion on which we find an opinion expressed is in his remark about the sacrilegious behaviour of Demetrius Poliorcetes, when he was initiated into the mysteries without observing the proper interval between one stage and the next: "This man does violence to all sacred ritual, to the rites both of the *mystae* and the *epoptae*" (since he did not wait to become an *epoptes* first before receiving full initiation as a *mystes*).[142] This expression of

[139] D.H. *Dein.* 3—Fg. 144.

[140] Did. col. 7, 23—26. Though this sentence is probably not a direct quotation, it doubtless reproduces many words and phrases of the original text. See note 112 above.

[141] Did. col. 1, 24—25. See p. 130 above.

[142] Fg. 148—Harp. *s.v.* ἀνεπόπτευτος· . . . ὁ μὴ ἐποπτεύσας. τί δὲ τὸ ἐποπτεῦσαι, δηλοῖ Φιλόχορος ἐν τῇ δεκάτῃ· " Τὰ ἱερὰ οὗτος ἀδικεῖ πάντα, τά τε μυστικὰ καὶ τὰ

opinion is interesting in view of the fact that Antigonus Gonatas is supposed to have ordered his execution; such outspoken disapproval of the' father of Antigonus doubtless did not aid his cause, when he was accused of intrigue with the Egyptian court. It also shows that his concern for correctness in sacred matters was a genuine one, if he was prepared to offend the feelings of such powerful persons as Demetrius Poliorcetes and Antigonus Gonatas. At the same time, the similarity of his interests to those of Demetrius of Phalerum suggests that, like Androtion, he sympathized with the "moderates" or even the oligarchs. As an Exegetes it is natural that he should be conservative and not disposed to welcome constitutional change unless there was precedent for it. But the fragments do not suggest that he was ready to proffer political opinions at all and give us no grounds for believing that his history represented the views of any particular party.

However inadequate our knowledge of the methods of Philochorus must be, we can still see fairly clearly the opinion that later writers held of him. As an historical authority, whether for the fifth or fourth century, he was convenient to use and evidently enjoyed a reputation for trustworthiness. On the other hand, nothing suggests that his *Atthis* had any remarkable quality or distinction as a literary work or that he had any particular insight into the history of the period. In their bald presentation of the facts the fragments recall the apparent impartiality of Thucydides. But mere fragments tell too little for us to know whether this appearance of impartiality disguised any deep thought or definite sympathies.

BIBLIOGRAPHY

The fragments

C. G. Lenz and C. G. Siebelis, *Philochori Atheniensis librorum fragmenta* (Leipzig, 1811).

C. and T. Müller, *Fragmenta Historicorum Graecorum* (*FHG*) 1.lxxxiv–v, lxxxviii–xl, 384–417; 4.646–48.

R. Stiehle, "Zu den Fragmenten der griech. Historiker," *Ph* 8 (1853) 638–43.

J. Strenge, *Quaestiones Philochoreae* (Diss. Göttingen, 1868).

Didymus, *Commentary on Demosthenes* (see Bibliography on Androtion).

W. Crönert, "Neue Lesungen des Didymospapyrus," *RhM* 62 (1907) 380–89.

Academicorum Philosophorum Index Herculensis, edited by S. Mekler (Berlin, 1902).

R. Reitzenstein, "Ein Bruchstück des Philochoros," *NGG*, phil.-hist. Klasse, 1906, 40–48.

A. Tresp, *Die Fragmente der griech. Kultschriftsteller, Religionsgesch. Versuche u. Vorarbeiten* 15 (Giessen 1914), 70–85.

ἐποπτικά." καὶ πάλιν· " Δημητρίῳ μὲν οὖν ἴδιόν τι ἐγένετο παρὰ τοὺς ἄλλους, τὸ μόνον ἅμα μυηθῆναι καὶ ἐποπτεῦσαι, καὶ τοὺς χρόνους τῆς τελετῆς τοὺς πατρίους μετακινηθῆναι."

Discussion

F. Albracht, *De Themistoclis Plutarchei fontibus* (Diss. Göttingen, 1873).

A. Böckh, "Ueber den Plan der Atthis des Philochoros," *Abh. Berlin•Akad.* phil.-hist. Klasse, 1832, reprinted in his *Gesammelte Kleine Schriften* 5.379–429.

P. Foucart, "Étude sur Didymos," *Mém. de l'Acad. des Inscriptions* 38 (1909), 27–218 (esp. 159–211).

W. Fricke, *Untersuchungen über die Quellen des Plutarchos im Nikias u. Alkibiades* (Diss. Leipzig, 1869).

G. Gilbert, "Die philochoreischen ὁμογάλακτες," *Jahrb. f. Philol.* 107 (1873) 44–48.

Id. "Die Quellen des plutarchischen Theseus," *Ph* 33 (1874) 46–66.

M. Lenchantin De Gubernatis, "Frammenti Didimei di Filocoro," *Aegyptus* 2 (1921) 23–32.

Id. "Nuovi frammenti di Filocoro," *RFIC* N.S. 10 (1932) 41–57.

R. Laqueur, *RE s.v.* "Philochoros."

W. Meiners, *Quaestiones ad scholia Aristophanica historica pertinentes, Diss. philol. Halenses* 11 (1890), 219–402 (esp. 336–72).

A. Philippi, *Commentatio de Philisto, Timaeo, Philochoro Plutarchi in Niciae vita auctoribus* (Giessen, 1874).

A. Roersch, "Étude sur Philochore," *MB* 1 (1897) 57–80, 137–57, 160–74.

K. Schenkl, *Bursian's Jahresb.* 34 (1883) 234–35.

C. G. Siebelis, *Observationes in locos quosdam Philochori difficiliores* (Programm, 1811). (Not accessible to me.)

C. Sintenis, "Zur Frage über das Attische Bürgerthum," *Ph* 5 (1850) 27–35.

F. Stähelin, "Die griech. Historikerfragmente bei Didymos," *Kl* 5 (1905) 55–71.

J. Strenge, *De Philochori operum catalogo qui exstat apud Suidam quaestio*, in *Festschr. for E. Curtius* (Göttingen, 1868).

F. Susemihl, *Gesch. der griech. Literatur in der Alexandrinerzeit* 1.594–99.

U. von Wilamowitz-Möllendorf, "Die Herkunft des Philochoros," *H* 20 (1885) 631.

J. H. Wright, "Did Philochorus quote the 'Aθ. Πολ. as Aristotle's?", *AJPh* 13 (1891) 310–18.

II. ISTER

Ister is generally reckoned as one of the Atthidographers, since he wrote a work on Attic antiquities which covers some of the same ground that earlier *Atthides* had covered, but in reality he stands quite apart from them. He was not an Athenian, very probably was not of free birth; and there is no suggestion that he ever took part in political life or held any priestly positions. There is no evidence that his *Atthis* dealt with historical times at all, since quotations from the thirteenth and fourteenth books are concerned with Theseus; and it is very doubtful if it ought to be called by that name (though it will be so called in this chapter for the sake of convenience). His connection with Callimachus, which links him with such men as Philostephanus and Hermippus, who collected antiquarian material, gives him a better title to belong among the scholars and grammarians of Alexandria than among the literary men of Athens. Some apology, therefore, seems to be needed for devoting space to him here.

Something must first be said about the attention he has received from earlier critics. Because his work on Attic questions is sometimes quoted under the title of A Collection or Collection of Atthides (Συναγωγή, 'Ατθίδων Συναγωγή), some scholars have believed that he summarized and quoted generously from earlier Atthides. This view, justified not so much by the fragments themselves as by his connection with Callimachean scholarship, led on to the belief that later authors, such as Plutarch, Pausanias, and Athenaeus, who quote from the Atthidographers, had not seen their works at first hand but derived their knowledge of them from Ister;[1] that, instead of mentioning Ister, they gave the name only of Cleidemus or Androtion or whatever author Ister quoted, not wishing their readers to know that their quotation was derived from an intermediate source. Such a theory cannot be disproved; we do, in fact, find Plutarch referring to Ister as recording the versions of other writers and the scholiast on Aristophanes speaks of him as "compiling the statements made by the historians" (τὰ παρὰ τοῖς συγγραφεῦσιν ἀναλεγόμενος).[2] There is, however, nothing to show that the work of Ister was more generally known and more accessible than that of these earlier writers; and there is no evidence at all that he enjoyed a greater reputation than they did.[3]

Even though we set aside the theory that he was the intermediate source through which later writers learnt about the Atthidographers, we must still grant that he quoted from them and borrowed from them and certainly dealt in similar fashion with some of the same mythological material. This seems a valid excuse for discussing his fragments briefly and for showing in what respects he differed from his predecessors. His apparent effort to collect a great mass of material about Attic mythology need not surprise us, since he is an Alexandrian writer. Completeness is likely to be his aim rather than elucidation or consistency, and this difference in purpose explains his need of so much space in dealing with mythical times.[4] No fragment suggests that he offered original

[1] This theory is most fully developed by M. Wellmann, De Istro Callimachio, who does not believe that Plutarch and Pausanias had read as widely as their quotations imply. Cf. also W. Graf Uxkull-Gyllenband, Plutarch und die griech. Biographie 69–76, who, however, confuses Ister and Philochorus, attributing an 'Ατθίδων Συναγωγή to the latter.

[2] Fg. 11, 12—Sch. Ar. Av. 1694, Plu. Thes. 34.

[3] Cf. Wellmann, op. cit. 33–35.

[4] A complete discussion of Ister should aim at considering his place in the development of Alexandrian scholarship. No attempt to do this can be made here.

explanations or improved versions of legends, such as would serve any patriotic purpose or justify the conduct of gods or heroes. Thus, as a neutral Alexandrian rather than a loyal Athenian, he stands apart from the Atthidographers properly so-called; this need not mean that he had acquired a scientific spirit or a critical faculty which the others lacked; but it may mean that he had a greater acquaintance with earlier and less familiar literature.

Suidas [5] says that he was a Cyrenean or a Macedonian or possibly a native of Paphos; and that he was a slave and friend of Callimachus. If he was a slave or of servile origin, the uncertainty about his native land is readily understood. It has been suggested that his title of Cyrenean is merely the result of his association with Callimachus and his title of Macedonian the result of his residence at Alexandria; but there is no such ready explanation of his connection with Paphos. It is possible, moreover, that he is the same person as "Ister the Callatian," from Callatis on the Danube, said to be the author of "an excellent book *On Tragedy*." [6] Several fragments show that our Ister—"the Callimachean," as Athenaeus calls him [7]—was interested in literary history. The *Life of Sophocles* quotes an Ister (without further qualification) for personal details about the dramatist; and it seems most likely that this is the same person who is quoted in the scholia on the *Oedipus Coloneus* for points of Attic topography and mythology. Suidas credits our author (the only Ister whom he mentions) with a book on *Lyric Poets;* [8] and we find the same authority cited for events in the life of Pindar, Xenophon, and Choerilus of Samos, as well as for Timaeus' nickname *Epitimaeus*. [9] There is, therefore, quite a good case for

---

[5] *S.v.* Ἴστρος, Μενάνδρου, Ἴστρου (ἢ Ἴστρου, Siebelis), Κυρηναῖος ἢ Μακεδών, συγγραφεύς, Καλλιμάχου δοῦλος καὶ γνώριμος. Ἕρμιππος δὲ αὐτόν φησι Πάφιον ἐν τῷ δευτέρῳ τῶν διαπρεψάντων ἐν παιδείᾳ δούλων. ἔγραψε δὲ πολλὰ καὶ καταλογάδην καὶ ποιητικῶς. It is uncertain whether this Hermippus is "the Callimachean," who wrote extensively on biography, or Hermippus of Berytus, who belongs to the age of Hadrian. Cf. Heibges in *RE s.v.* "Hermippos" (6) & (8).

[6] St. Byz. *s.v.* Κάλατις, πολίχνιον ἐν τῇ παραλίᾳ τοῦ Πόντου, ἀφ' οὗ Ἴστρος Καλατιανός, περὶ τραγῳδίας γράψας καλὸν βιβλίον.

[7] 6.272B; 10.478B.

[8] *Fg.* 50—Suid. *s.v.* Φρῦνις· . . . Ἴστρος δ' ἐν τοῖς ἐπιγραφομένοις Μελοποιοῖς τὸν Φρῦνιν Λέσβιόν φησι, Κάνωπος υἱόν. Susemihl, *Gesch. der griech. Lit. in der Alexandrinerzeit*, 1.512, 625, is inclined to believe that the references in the *Life of Sophocles* are to our Ister, but that the Μελοποιοί was written by Ister of Callatis, a different person, about whose date nothing can be known.

[9] *Pindari Vita Ambros.* (cited by Jacoby, *RE* 9.2282, not in *FHG*); D.L. 2.59—Fg. 24; Ath. 8.345D, 6.272B (see Jacoby, *loc. cit.*).

identifying Ister "the Callimachean" with Ister of Callatis, and Crusius actually suggests that, as a slave, he was called Ister because his native place was on the River Danube.[10]

There is nothing to establish a definite date for our author except his connection with Callimachus and the fact that he was sternly criticized by Polemon of Ilium, the Periegetes; Polemon, who is said to have "flourished" in the early part of the second century B.C., in the reign of Ptolemy V, declared that Ister deserved to be thrown into the River Ister.[11]   The keen interest of Callimachus in Attic myths, and indeed in aetiological legends generally, renders it probable that he employed Ister in collecting the material which he incorporated in his *Aetia* and his *Hecale*.

There is some difficulty in establishing the list of Ister's works. Nothing is known of the poetry which Suidas says that he wrote.[12] It is possible, however, that his discussion of *Colonies of the Egyptians* (cited in two fragments) [13] is simply a digression in his *Atthis*. The relations between Athenians and Egyptians in early times and their mutual borrowings had provided a favourite topic for discussion ever since the time of Herodotus; Philochorus recorded the opinion that Cecrops was an Egyptian, whereas Phanodemus argued that the Athenians were really "fathers of the Saïtes." [14] In a later echo of this controversy, Diodorus remarks that "the Egyptians say their forefathers sent out many colonies to different parts of the world"; but he decides not to record their claims in detail because "no sure proof of their accuracy is available and no trustworthy historian bears witness to them." [15]   It is arguable that Diodorus is here referring, in rather slighting terms, to Ister, and that this author in his *Atthis* developed the theories propounded by earlier Atthidographers.[16]   On the other hand, since Athenaeus credits Ister with a *Ptolemaïs*,[17] he may well have written more than one separate work on an Egyptian theme.

[10] *Sitz.-Ber. der Münchener Akad.* (1905) 794.

[11] Ath. 9.387F.

[12] See note 5 above.   Suidas makes a similar remark about Hellanicus (*s.v.* Ἑλλάνικος): συνεγράψατο δὲ πλεῖστα πεζῶς τε καὶ ποιητικῶς.

[13] Fg. 39, 40.

[14] Philoch. fg. 10; Phanod. fg. 7.

[15] 1.28.1; 29.5.   His argument in these two chapters shows that he is familiar with the views of Phanodemus and Philochorus.

[16] Cf. Wellmann, *op. cit.* 12.

[17] 10.478B—Fg. 38: Ἴστρος ὁ Καλλιμάχειος ἐν πρώτῳ Πτολεμαΐδος, τῆς ἐν Αἰγύπτῳ πόλεως, γράφει οὕτως· " Κυλίκων Κονωνίων ζεῦγος καὶ θηρικλείων χρυσοκλύστων ζεῦγος."

Scarcely anything can be discovered about his other minor works, though their titles suggest similar books to those written by the earlier Atthidographers. They included a grammatical work ('Ἀττικαὶ λέξεις),[18] some works on local history and customs (*Argolica, Eliaca*),[19] discussions of religious matters ('Ἀπόλλωνος ἐπιφάνειαι, Συναγωγὴ τῶν Κρητικῶν θυσιῶν),[20] and a work on *Contests;* [21] the title *Notes* ('Ὑπομνήματα) may or may not indicate some special work.[22]

As regards the title of his major work on Attic affairs, the confusion seems at first almost hopeless. It is never called an *Atthis*, though sometimes a *Collection of Atthides* ('Ἀτθίδες, 'Ατθίδος or 'Ατθίδων Συναγωγή); on other occasions the title appears as *Attic Collections* or simply *Attic affairs* ('Ἀττικαὶ Συναγωγαί, 'Ἀττικά). Further difficulty is caused by the references to Ἄτακτα, a title which Wellmann would identify with the miscellaneous *Notes* ('Ὑπομνήματα).[23] The probable explanation is that Ἄτακτα and Συναγωγή are alternative titles, since in the two fragments referring to the Eumolpidae at Eleusis the *Synagoge* is cited in one instance and the *Atakta* in the other.[24] The former title is used by Antigonus of Carystus and Antoninus Liberalis for a collection of poems or stories, though its use by the Peripatetics and medical writers for their encyclopaedic works is better known.[25] A parallel for a similar use of *Atakta* is Euphorion's *Mopsopia or Atakta*, the meaning of which is

---

[18] Fg. 53–55.

[19] Fg. 43–46. Cf. also Sch. Pi. O. 6.55 (cited by Jacoby, *loc. cit.*).

[20] Fg. 33–37, 47.

[21] Περὶ ἰδιότητος ἀγώνων (Fg. 48). Περὶ ῾Ηλίου ἀγώνων (Fg. 60b) is probably a subdivision of this work.

[22] Fg. 52. Possibly the miscellaneous mythological fragments 56–59, 61–65, should be assigned to this work; also the new fragments cited by Jacoby: Sch. Townley *Il.* 19.119, and *POxy.* 2.221, col. 6.28–30.

[23] *De Istro Callimachio* 5–7. Cf. Susemihl, *op. cit.* 623.

[24] Fg. 20—Sch. Lyc. 1328: Εὔμολπος, οὐχ ὁ Θρᾷξ κατὰ τὸν Ἴστρον ἐν τῇ Συναγωγῇ, ἀλλ' ὁ θεὶς τὰ μυστήρια, ἐκέλευ·τε ξένους μὴ μυεῖσθαι, ἐλθόντος δὲ τοῦ ῾Ηρακλέους ἐν 'Ελευσῖνι καὶ θέλοντος μυεῖσθαι, τὸν μὲν τοῦ Εὐμόλπου νόμον φυλάττοντες, θέλοντες δὲ τὸν κοινὸν εὐεργέτην ῾Ηρακλέα θεραπεῦσαι, οἱ 'Ελευσίνιοι ἐπ' αὐτῷ τὰ μικρὰ ἐποιήσαντο μυστήρια· οἱ δὲ μυόμενοι μυρσίνῃ ἐστέφοντο. Fg. 21—Sch. Soph. *OC* 1053: ζητεῖται τί δήποτε οἱ Εὐμολπίδαι τῶν τελετῶν ἐξάρχουσι, ξένοι ὄντες. εἴποι δ' ἄν τις, ὅτι ἀξιοῦσιν ἔνιοι, πρῶτον Εὔμολπον ποιῆσαι τὸν Δηϊόπης τῆς Τριπτολέμου τὰ ἐν 'Ελευσῖνι μυστήρια, καὶ οὐ τὸν Θρᾷκα, καὶ τοῦτο ἱστορεῖν Ἴστρον ἐν τῷ περὶ (Elmsley, πέμπτῳ) τῶν ἀτάκτων.

[25] Plu. *Mus.* 3.1131f; 5.1132f; Antig. Caryst 26 (32); Gal. (ed. Kühn) 12, p. 836; Orib. *Collectiones*, 1. *Proem.* Cf. H. Étienne, *Thesaurus Linguae Graecae, s.v. συναγωγή,* section entitled *Collectio, de libris aliisque scriptis.*

explained by Suidas:[26] it was called *Atakta* "because it contained various stories (συμμιγεῖς ἱστορίας), *Mopsopia* because Attica was formerly called by that name and the story of the poem is concerned with Attica." It seems likely, then, that the two titles Ἄτακτα and Συναγωγή, or in longer form Ἄτακτα Ἀττικά, Ἀτθίδων Συναγωγή, were used to describe Ister's long and perhaps rather loosely organized work on various subjects connected with Attic history and Attic customs. In that case, the title does not imply any intention on Ister's part to collect or summarize the work of earlier Atthidographers.

One must look to the fragments, however, rather than to the uncertain evidence of the titles for information about the character of his work. There are two fragments which suggest that, instead of giving a single definitive account of a myth, he offered more than one alternative. Harpocration quotes a sentence discussing the behaviour of Erechtheus after his daughter Procris had been killed by Cephalus: "And some say that Erechtheus is represented with his spear fixed in the ground over the grave as pledging his spear and signifying his distress, because it was the custom for relatives of the deceased to take action against murderers in this manner."[27] So also Plutarch in his *Theseus* writes that "Ister gives a peculiar and entirely different account of Aethra in the thirteenth book of his *Attica;* he says that, according to some accounts (ἐνίων λεγόντων), Alexander (i.e. Paris) was beaten in battle in Thessaly beside the Spercheius by Achilles and Patroclus, and that Hector captured and sacked the city of the Troezenians and carried off Aethra who was left behind there. But this account (Plutarch adds) is entirely unreasonable."[28] Apart from these fragments which show that Ister liked to record and perhaps to discuss the versions of other writers, there are two citations which show him adopting the explanations given by Philochorus; and another fragment tells us that he followed Hellanicus and Androtion in their accounts of the institution of the Panathenaea by Erichthonius.[29]

[26] *S.v.* Εὐφορίων. Cf. also C. Cessi, "Euphorionea," *RFIC* 43 (1915) 278–92 and F. Marx, *Lucilius,* 1, Prolegomena xiv, who suggests *Atakta* as the equivalent of *carmina per saturam.* The Ἄτακτοι γλῶσσαι of Philetas and the Ἄτακτοι λόγοι of Simonides seem to offer no parallel.

[27] Fg. 19—Harp. *s.v.* ἐπενεγκεῖν δόρυ.

[28] Fg. 12—Plu. *Thes.* 34.

[29] Ister fg. 1, Philoch. fg. 157—Suid. *s.v.* Τιτανίδα γῆν, οἱ μὲν τὴν πᾶσαν, οἱ δὲ τὴν Ἀττικήν, ἀπὸ Τιτηνίου, ἑνὸς τῶν Τιτάνων ἀρχαιοτέρου, οἰκήσαντος περὶ Μαραθῶνα, ὃς μόνος οὐκ ἐστράτευσεν ἐπὶ τοὺς θεούς, ὡς Φιλόχορος ἐν Τετραπόλει, Ἴστρος δ' ἐν α' Ἀττικῶν.

Plutarch found the story about Aethra in the thirteenth book and Athenaeus takes another reference to the legend of Theseus from the fourteenth.[30] Evidently, therefore, Ister treated the legendary period in great detail and there is no evidence that he ever came down as far as historical times. Such an extensive treatment of mythical times would make it possible for him to record all manner of different versions, irrespective of whether they were credible or reflected credit on the gods and heroes involved. We learn from Athenaeus that he "gave a list of the women of Theseus," saying that "Theseus won some of them by love, others by violence, and others by lawful marriage; he won Helen, Ariadne, Hippolyte, and the daughter of Cercyon and Sinis by violence; but he married Meliboea, the mother of Ajax." [31] This manner of treating Theseus seems very different from the attempt of Philochorus to idealize him as a law-abiding national hero.[32]

On the other hand, some of the usual characteristics of *Atthides* are well represented in the fragments. His interest in *aetia* and etymologies is shown by his account of Titenius, the Titan of Marathon, from whom Attica obtained its name of *Titanis;* by his explanations of Ταυροπόλος as a name of Artemis and Ὁμολώιος as a name of Zeus; and by his connection of the name of the month Anthesterion with the flowers that bloomed in that season.[33] He wrote about the meaning and the origin of other festivals: the torch festivals, the Theoinia, the Panathenaea, the Oschophoria, and the procession in honour of Erse, the daughter of Cecrops.[34] He mentioned the plants specially sacred to Demeter and the wreaths made from them worn by her priests and priestesses; and he evidently gave some account of the Eumenides.[35] He distinguished the Thracian Eumolpus from the founder of the Eleusinian mysteries; and he described the initiation of Heracles into the "lesser mysteries." [36] His religious interests, it seems, are very similar to those

For the text see section on Philochorus, note 18. Cf. Ister fg. 26, Philochorus, new frag. no. 3, Reitzenstein, *NGG* (1906) 41–42. Ister fg. 7—Harp. *s.v.* Παναθήναια· . . . ἤγαγε δὲ τὴν ἑορτὴν πρῶτος Ἐριχθόνιος ὁ Ἡφαίστου, καθά φασιν Ἑλλάνικός τε καὶ Ἀνδροτίων, ἑκάτερος ἐν πρώτῃ Ἀτθίδος. πρὸ τούτου Ἀθήναια ἐκαλεῖτο, ὡς δεδήλωκεν Ἴστρος ἐν τρίτῃ τῶν Ἀττικῶν.

[30] Ath. 13. 557A—Fg. 14.
[31] See note 30 above. Cf. Plu. *Thes.* 29.
[32] See section on Philochorus, pp. 114–15 above.
[33] Fg. 1 (see note 29 above), 8 (cf. Phanod. fg. 10), 10, 28.
[34] Fg. 3, 4, 5, 7, 13, 17. Cf. Philoch. fg. 14, 44.
[35] Fg. 25, 9.
[36] Fg. 20, 21. See note 24 above.

of the earlier Atthidographers. He was also an assiduous collector of oracular responses, like Philochorus and Herodotus.[37]

The fragments contain a number of remarks about Attic topography. In the *Oedipus Coloneus* of Sophocles, after Theseus and a band of Athenians have set out in pursuit of the Thebans who have seized Antigone and Ismene, the chorus sings of the places they will pass on their route. In commenting on this passage, the scholiast seeks explanation in Book I of Ister's *Atakta* and quotes a sentence which seems to come from a description of the boundaries of the deme Oea.[38] Harpocration also quotes the *Atakton* for a distinction between the demes Paeaneis and Paeonides.[39] Since the scholiast cites the first book, it is hardly possible that this description of the demes was given in connection with his treatment of Cleisthenes.[40] It seems more likely that Ister gave a general topographical description of Attica before starting his narrative. Such a procedure would not be contrary to the practice of earlier historians, but there is no evidence that any of the Atthidographers followed it.[41]

On another occasion the scholiast on Sophocles says that Ister gave the number of the μόρια, the sacred olives.[42] But his trust-

[37] Plu. *Pyth. Orac.* 403E—Philoch. fg. 195: μυρίους τοίνυν καὶ Ἡροδότου καὶ Φιλοχόρου καὶ Ἴστρου, τῶν μάλιστα τὰς ἐμμέτρους μαντείας φιλοτιμηθέντων συναγαγεῖν, ἄνευ μέτρου χρησμοὺς γεγραφότων, Θεόπομπος οὐδενὸς ἧττον ἀνθρώπων ἐσπουδακὼς περὶ τὸ χρηστήριον ἰσχυρῶς ἐπιτετίμηκε τοῖς μὴ νομίζουσι κατὰ τὸν τότε χρόνον ἔμμετρα τὴν Πυθίαν θεσπίζειν.

[38] The particular lines in question are *OC* 1059–1061:

> ἤ που τὸν ἐφέσπερον
> πέτρας νιφάδος πελῶσ'
> Οἰατίδος εἰς νομόν.

The scholiast comments: τὸν ἐφέσπερον. τὸν Αἰγιάλεών φησι. καὶ γὰρ τοῦτο ἐπ' ἐσχάτων ἐστὶ τοῦ δήμου τούτου. καταλέγουσι δὲ χωρία, παρ' ἃ μάλιστα εἰκάζουσι τὴν συμβολὴν γενέσθαι τοῖς περὶ τὸν Κρέοντα καὶ Θησέα. πέτρας δὲ νιφάδος ἂν εἴη λέγων (sc. ὁ Σοφοκλῆς) τὴν οὕτω λεγομένην λείαν πέτραν ἢ τὸν Αἰγιάλεων λόφον, ἃ δὴ περιχώριά φασιν εἶναι, καθάπερ Ἴστρος ἐν τῇ πρώτῃ τῶν Ἀτάκτων ἱστορεῖ, οὕτως· " Ἀπὸ δὲ τῆς χαράδρας ἐπὶ μὲν λείαν πέτραν." καὶ μετ' ὀλίγα· " Ἀπὸ τούτου δὲ ἕως Κολωνοῦ παρὰ τὸν Χαλκοῦν προσαγορευόμενον ὅθεν πρὸς τὸν Κηφισὸν ἕως τῆς μυστικῆς εἰσόδου εἰς Ἐλευσῖνα. ἀπὸ ταύτης δὲ βαδιζόντων εἰς Ἐλευσῖνα τὰ ἐπαρίστερα μέχρι τοῦ λόφου τοῦ πρὸς ἀνατολὴν τοῦ Αἰγιάλεω."

[39] Fg. 31.

[40] See above pp. 117–18 and note 72.

[41] Jebb, *Sophocles, Oedipus Coloneus*, Intro. xxxvi, is hardly justified in saying that Ister was one of the chief authorities on Attic topography in the later Alexandrian age. He thinks that the passage from which the scholiast quotes is an "itinerary of Attica"; but this explanation ignores τὰ ἐπαρίστερα in the last sentence, which Jebb does not quote in full.

[42] Sch. Soph. *OC* 697—Fg. 27.

worthiness as a topographical guide seems rather questionable, in view of his remarks about the well called Clepsydra on the Acropolis. "Clepsydra is a well on the Acropolis," writes the scholiast on Aristophanes, "which Ister mentions in his twelfth book, collecting the statements in the historians; and it is so called because, when the Etesian winds begin to blow, it fills, and when they cease the water sinks, just like the Nile and also the well at Delos. He says that a *phiale* stained with blood fell into it and was seen again in the Bay of Phalerum, twenty stades away; and they say that the well is enormously deep and that its water is salt." [43] If the scholiast is quoting accurately, this fragment seems to show that Ister, as a dweller in Alexandria, contented himself with recording statements about the Athenian Acropolis which he did not properly understand; a visit to Athens would have taught him that the salt spring on the Acropolis was not Clepsydra, but Poseidon's *Thalassa Erechtheis* in the Erechtheum.[44]

The fragments show that Ister's work, whatever its exact title, contained several of the characteristics common to *Atthides*. But the task of reconstructing its arrangement is an impossible one. There are no fragments referring to events of historical times and the solitary reference to the sixteenth book [45] is not enough to tell us what topics he treated after he had finished with Theseus. We are obliged to say, therefore, that, although it resembles the work of the Atthidographers in some ways, it seems not to correspond to the scheme of any *Atthis* known to us.

BIBLIOGRAPHY

C. G. Lenz and C. G. Siebelis, *Phanodemi, Demonis, Clitodemi atque Istri* Ἀτθίδων *et reliquorum librorum fragmenta* (Leipzig 1812) 51–80.

C. and T. Müller, *Fragmenta historicorum Graecorum (FHG)* 1.lxxxv, xc, 418–27; 4.648.

F. Jacoby, *RE s.v.* "Istros" (9).

O. Crusius, "Zur Beurteilung des Istros und der Atthidographen," *Sitz.-Ber. der Münchener Akad.* 1905, 793–99.

G. Gilbert, "Die Quellen des Plutarchischen Theseus" (see bibliography of Philochorus).

A. Goebel, "Nova historicorum Graecorum fragmenta," *Jahrb. f. class. Philologie* 93 (1866) 162–66.

F. Susemihl, *Gesch. der griech. Lit. in der Alexandrinerzeit*, 1.622–25.

M. Wellmann, *De Istro Callimachio* (Diss. Greifswald, 1886).

[43] Fg. 11—Sch. Ar. *Av.* 1694.

[44] Cf. Judeich, *Topographie von Athen* 259, 280.

[45] Fg. 16—Harp. *s.v.* Τραπεζοφόρος· Λυκοῦργος ἐν τῷ Περὶ τῆς ἱερείας ὅτι ἱερωσύνης ὄνομά ἐστιν ἡ τραπεζοφόρος· ὅτι αὐτή (leg. αὕτη) τε καὶ ἡ Κοσμὼ συνδιέπουσι πάντα τὰ τῆς Ἀθηνᾶς ἱερεῖα, αὐτός τε ὁ ῥήτωρ ἐν τῷ αὐτῷ λόγῳ δεδήλωκε καὶ Ἴστρος ἐν ἕκτῃ καὶ δεκάτῃ τῶν Ἀττικῶν συναγωγῶν.

CHAPTER VII

THE ATTHIS TRADITION

In previous chapters the special characteristics of individual historians have been investigated, so far as they are revealed in complete works or in fragments. Whilst the relation of each writer to other Attic historians has been borne in mind, no author's place in the literary succession can be properly established until the work of both his predecessors and his successors has been examined. There is, indeed, a certain danger even in assuming that such a succession existed, since it is not fair to prejudge any literary work by taking it for granted that it conforms or ought to conform to a norm of tradition. The aim, therefore, of the previous chapters has been to present the evidence and draw such conclusions as followed from the evidence examined at the time. The task of summing up the evidence and presenting the conclusions which emerge from it when it is taken as a whole has been reserved for this final chapter.

Modern historians sometimes speak of the Atthis tradition in the sense of an established historical tradition; their implication is that the Atthidographers collectively established and perpetuated certain views about Athenian history, which came to be accepted as traditional; and that we should be able to reconstruct this tradition in great part if we possessed the full text of one *Atthis*. But the fragments of the Atthidographers, as they have been examined in the preceding chapters, have given no ground for believing in any such traditions of historical opinion. On the contrary, there is abundant evidence that these writers disagreed with one another on a number of points. If there was such a thing in Athens as a body of historical tradition, generally accepted by people of conservative tendencies, the evidence for it must be sought elsewhere; any attempt to conjecture the current opinions of the general public on historical questions must rest on evidence sought from another quarter.[1] The Atthis tradition, which forms the subject of the present chapter, is not an historical but a literary tradition.

[1] Especially on evidence sought from the Attic orators, who sometimes appeal to the historical knowledge of their auditors.

The aim of the following discussion is to show that there was a continuous literary tradition, which the local historians kept alive with
a certain degree of progress and development. No tradition can
remain alive if those who inherit it from one generation to another
make no contributions to it; and it certainly is not likely that
Atthides would continue to command respect, if each writer regarded
it as his principal task to pass on a body of historical information
that his precedessors had already collected.

The *Atthides* provide us with examples of Attic local history,
but we have scarcely any information about the writing of local
history in other cities from the fifth to the third century.[2] The
fragments of Antiochus offer some indication of what was being
done in Syracuse in the fifth century, but later on the reputation
of Timaeus was such as to overshadow his less rhetorical contemporaries; we know of no representatives of the more sober style in
Sicily who rival him even to the extent that Androtion and Philochorus rival Ephorus and Theopompus. But so far as the Attic
writers are concerned, we can be fairly certain from the evidence of
the fragments that they were not prone to rhetorical devices or
moral reflections. The *Atthides* seem to have been distinguished by
their conservatism and respect for traditional religion; and this is
not surprising when one remembers that some of their authors held
priestly offices. We know that in some respects the Atthidographers
followed in the footsteps of the earlier Ionian logographers, who,
according to Dionysius of Halicarnassus, tried to "bring to the
knowledge of the public the written records preserved in temples
or in secular buildings in the form in which they found them, neither
adding nor taking away anything."[3] It seems worth while to
discuss their loyalty to the Ionian literary tradition more fully.

The earlier Ionian logographers linked together the discussion
of myths, genealogies, and local history. Herodotus and Thucydides, on the other hand, tried to narrow their field and disclaimed
any desire to unearth the truth about the more distant past. They
were not, however, entirely successful in this attempt to limit their
theme; from time to time they indulged in mythological digressions,
since they found it impossible, as also did Ephorus, to divorce the

---

[2] When the third volume of Jacoby's *FGrH*, containing the fragments of *Ethno-
graphie* and *Horographie*, is completed, the extent of our knowledge about Greek local
historians will be more easily recognized. Meanwhile see Jacoby, *Kl* 9 (1909) 109–21.

[3] D.H. *Th.* 5.

immediate past completely from the more remote past in which the traditions of the Hellenic peoples had been founded. Whatever claims the Atthidographers may have made, they certainly did not confine themselves to the history of more recent events; they began their history of Athens at the very beginning and faced the question whether the Athenians were autochthonous or of Egyptian origin.

Thucydides also claimed that his history would be a "possession for ever," a record of events which would have permanent value for later generations, rather than a *tour de force* which would give pleasure for the moment. This claim was undoubtedly felt as a challenge by the writers of the fourth century. The mere recording of events, with no seasoning of any kind, was naturally enough an unpalatable form of literary composition. The pupils of Isocrates, therefore, stressed the moral lessons to be learnt from history and did their best to narrow the breach between history, oratory, and philosophy. Even the earlier Atthidographers seem to have linked historical narrative with religious discussion; and as the moralistic approach to literature became more general through the fourth century, Philochorus tried to make his religious discussion conform to the ethical interests of his readers. Meanwhile, the Peripatetics were starting to write biographies in which the ethical interest played a prominent part. As professional philosophers, the Peripatetics could claim a certain authority in matters of ethics; but an Exegetes like Philochorus also had a claim to authority. Unlike the theoretical philosophers and rhetoricians, unlike Ephorus but like Thucydides and Xenophon, some of the Atthidographers could claim experience of the public life of the city, whether in politics (as Androtion) or in sacred office (as Cleidemus and Philochorus). Furthermore, if a politician could claim special understanding of the political history of Athens, an Exegetes could claim special knowledge of Athenian national myths and their significance for religious life.

Dionysius of Halicarnassus finds fault with the earlier logographers, because they wrote either "Greek histories" or "barbarian histories," but made no attempt to describe the common fortunes of Greeks and barbarians as Herodotus did. Here again the Atthidographers, following the lead of Thucydides, preferred the more restricted field. *Hellenica* as a theme had become unwieldy for a writer who was concerned to be accurate. Accordingly, when the Atthidographers stray beyond the bounds of Attic history, we find that they still restrict their theme κατὰ πόλεις, writing *Eliaca, Deliaca,* or *Epirotica,* rather than *Hellenica.* One can explain this preference for a restricted theme not only by the forbidding bulk of material relevant for a general history of the Hellenic world, but also by the individualist political outlook of the fourth century. The comprehensive work of Herodotus, as Jacoby points out in an excellent essay,[4] was to some extent the result of the panhellenic feeling which the Persian Wars inspired; it is the common enmity of the Greek states to Persia which gives unity to his work. In the fourth century, again, the attempt of Isocrates to revive this pan-hellenic feeling may be held responsible for Ephorus' plan in writing a comprehensive history. His Ἐπιχώριος λόγος about his native Cyme is really more in keeping with the political spirit of the age.

The Atthidographers, however, restricted themselves in a similar way in dealing with mythology. Here they followed the lead of Hellanicus, who, instead of writing a comprehensive Ἡρωολογία like Hecataeus, dealt with the various heroic families separately in works like his *Phoronis, Deucalioneia,* and *Asopis.* The material of Greek mythology had become too bulky for comprehensive treatment, and awaited the selective hand of the Hellenistic compilers. By confining themselves to Attic myths the Atthidographers set themselves a less impossible task. But in time even Attic mythology increased to an unwieldy size, so that when Ister set out to give a complete account of it he had enough material to fill at least fourteen books. The preceding discussion of the fragments has shown that the Atthidographers did not merely hand on the old myths as they found them. Each one added something of his own, some new interpretation or some new incident; Philochorus was just as ready to make a new contribution as Hellanicus had been. In this respect, certainly, they were not conservative. But individual interpretation of myths was not a new thing; it had been in fashion ever since the time of the early logographers and was part of the Ionian tradition of ἱστορίη. Thus the Atthidographers could claim well-established precedent for exercising their imagination and ingenuity in this field. Sometimes they even ventured outside the limits of Attic legend, as Philochorus did in discussing and rationalizing the legend of Dionysus.[5]

[4] "Griechische Geschichtschreibung," *Die Antike* 2.1–29, esp. 11–15.
[5] Cf. Chap. 6, p. 113 above.

It would be most interesting to know for certain what particular tendency this rationalistic interpretation took in different periods. The fragments offer enough examples to make it evident that each Atthidographer indulged in it, but conclusions about its development from Hellanicus to Philochorus can be only tentative. Hellanicus explained the Homeric tale of Achilles' fight with the River Scamander by saying that the river overflowed its banks, apparently because of rainfall in the mountains,[6] but he did not delete all miracles from heroic legend; he retained incidents just as startling as the fight of the river with Achilles, such as the growth of men from the dragon's teeth, and did not deny that gods could take part directly in the affairs of men, as Poseidon and Apollo built the walls of Troy for Laomedon.[7] His books contained enough "marvels" (θαύματα) to call down the scorn of Strabo, who said that "one might as well believe Hesiod and Homer and the tragic poets, as Ctesias and Herodotus and Hellanicus and others of their kind." [8]

Plutarch records how Cleidemus altered the tale of Theseus' adventures at Cnossus; [9] how he assumed the existence of some international law regulating the sailing of ships on the sea. No instance of rationalism in treating other myths is recorded from his work, nor from that of Phanodemus, although Phanodemus altered the story of Agamemnon's sacrifice at Aulis, saying that Artemis substituted a bear, not a stag, for Iphigeneia.[10] Androtion seems to have rationalized the tale of the dragon's teeth, explaining the term *Spartoi* on the ground that the companions of Cadmus were "scattered wanderers" (σποράδες).[11] There is not so much evidence for the rationalistic methods of the earlier Atthidographers as for Philochorus, who seems to have removed all supernatural elements from the tale of Theseus and to have denied that he ever fought against any of the gods. The legend of Theseus, however, is repre-

---

[6] F.28. F. and T. are used to denote fragments and *testimonia* in Jacoby's *FGrH*. Fg. for those in Müller's *FHG*.

[7] F.1, 26.

[8] Str. 11.6.3—T. 24.

[9] Fg.5—Plu. *Thes.* 19.

[10] Fg. 10.

[11] Fg. 28, 29. Cf. the etymology of Αἰολεῖς from αἰολοι ("a motley crew") given by Sch. Lyc. 1374, which may be taken from Hellanicus (see Jacoby's note on Hellanicus F.32). A different account of the dragon's teeth by Androtion seems to be implied in Fg. 37—Sch. Lyc. 495: ὁ Αἰγεὺς Ἀθηναῖος καὶ γηγενὴς ἀπὸ Ἐρεχθέως. τινὲς δὲ καὶ τοῦτον ἕνα λέγουσι τῶν ἀναδοθέντων ἐκ τῶν ὀδόντων τοῦ δράκοντος τοῦ ἐν Θήβαις, ὡς καὶ Ἀνδροτίων.

sented by fragments from five different Atthidographers; hence, instead of attempting a comparison of their rationalistic methods in general, it seems best to confine the argument to their treatment of this legend.

The fragments of Hellanicus show no traces of rationalism here.[12] According to his account, Minos deliberately selects Theseus as one of the young Athenians to be sent to Crete, and an agreement has been made between Minos and the Athenians that the youths are to sail unarmed, so that their chances of killing the Minotaur are very slight indeed.   The founding of the Isthmian games, the expedition against the Amazons and the fight against them in Attica, and the abduction of Helen were all included in his account of Theseus. Plutarch [13] points out that some writers, since they considered the tale of his abduction of Helen a libel against him, tried to explain away his connection with the episode, saying that he merely consented to guard her after Idas and Lynceus had carried her off or that Tyndareus himself had entrusted her to him for safe keeping. Hellanicus shows no such anxiety to preserve Theseus' good name, but is more concerned to clear up the chronological difficulty, since Theseus is a generation older than Helen; he is obliged to make Helen a little girl seven years old carried off by Theseus when he is over fifty.   As for the rape of Persephone, he apparently followed the old version, according to which Theseus accompanied Peirithous on his journey to the lower world;[14] Plutarch prefers the later, rationalized version, that makes Persephone a daughter of Acdoneus, king of the Molossians, and Cerberus an ordinary fierce dog with no more than one head.[15]   It appears, therefore, that in Hellanicus' account Theseus has taken on the characteristics of an Athenian Heracles, without losing the more barbaric features of that hero; he founds games, just as Heracles did, and helps to civilize the world by fighting against the Amazons and killing the Minotaur, but he is also, like Heracles, ready to carry off women and defy

[12] See Chap. 1, pp. 18–19 for references.   H. Herter, "Theseus der Athener," *RhM* 88 (1939) 244–286, 289–326, wants to believe that Athenian national pride (in the 5th century) rejected the story that Theseus faithlessly abandoned Ariadne, regarding it as a slur on his character (p. 262).   He also thinks that "die Ethisierung der Theseusgestalt ist spätestens in der Peisistratidenzeit angebahnt worden" (p. 312).   But, though his articles are amply supplied with references on other points, he can quote no authority for these statements.

[13] *Thes.* 31.

[14] F.134.

[15] *Thes.* 31.

the powers of the lower world. Hellanicus, evidently, saw no reason to reject the stories that represented him as a woman hunter, which Ister afterwards attempted to catalogue; [16] his account shows no trace either of rationalism or idealization.

Thucydides was interested only in the political achievements of Theseus, but Plutarch finds the Atthidographers a fruitful source of information for the tales of his adventures. He calls Cleidemus' account of the expedition to Crete "individual and remarkable." The whole passage has been quoted in the discussion of Cleidemus [17] and need not be repeated here. Most significant is the fact that Minos is not reigning in Crete; in his pursuit of Daedalus over the sea he had been driven out of his course and died in Sicily. His successor, Deucalion, demands of the Athenians that they surrender Daedalus, but Theseus secretly builds a fleet and unexpectedly descends upon Cnossus, where he kills Deucalion in a battle at the gate of the labyrinth; then he makes a truce with Ariadne, who rules in Deucalion's place, and recovers the Athenian captives. This account shows a high degree of rationalism. The difficulty of reconciling the legend of Minos the just lawgiver with Minos the cruel tyrant is evaded, [18] and the tale of Theseus' love for Ariadne is not mentioned; perhaps, indeed, it is ignored because it reflects discredit on Theseus. Another side of Theseus' character is developed: his ability as a shrewd statesman and general. And the whole setting of the story is not heroic at all, with its assumption of "a general Hellenic decree" regulating the sailing of the seas in ships of war. In such a setting the Minotaur has no place and Cleidemus seems to have dispensed with it.

In the story of the battle with the Amazons at the foot of the Acropolis Plutarch emphasizes how Cleidemus gave a detailed account, [19] making the left wing of the Amazons "wheel towards the place now called the *Amazoneion*" and their right wing come up against the Pnyx, whilst the Athenians attacked them from the Hill of the Muses, and so on. Here again Cleidemus shows the skill of Theseus as a general, and "in the fourth month a truce is arranged through Hippolyte," all in proper statesmanlike, civilized

[16] Fg. 14.

[17] See Chap. 4, pp. 65–66 above.

[18] Cf. Ephor. F.147—Str. 10.4.8: ὡς δ' εἴρηκεν Ἔφορος, ζηλωτὴς ὁ Μίνως ἀρχαίου τινὸς Ῥαδαμάνθυος, δικαιοτάτου ἀνδρός, ὁμωνύμου τοῦ ἀδελφοῦ αὐτοῦ κ.τ.λ.

[19] Cleidem. Fg. 6—Plu. *Thes*. 26: ἱστορεῖ δὲ Κλείδημος, ἐξακριβοῦν τὰ καθ' ἕκαστα βουλόμενος.

fashion.   He has probably refashioned the whole legend of Theseus
and changed him from a rugged hero into the model of an Athenian
soldier-stateman.   It is a great pity that no further fragments of his
version of the legend survive.

The fragments of Phanodemus and Androtion contain nothing
relating to Theseus, but those of Demon and Philochorus show a
similar kind of rationalism, though the details are quite different.
Philochorus' treatment of Theseus has already been discussed,[20]
and only some of the points need be repeated here.   Both he and
Demon explained away the Minotaur; instead of a monster they
substituted Taurus, a general of Minos.   Demon follows Cleidemus
in having a battle, but it is a sea fight (a rarity for heroic times),
in which this Taurus is killed.   Philochorus has a much more
elaborate story.   In his account, as Plutarch gives it, nothing is
said of a battle.   His story, so far as it is preserved, is that games
were given at Cnossus in honour of the dead Androgeos (about whom
Melesagoras also had something to say),[21] in which the Athenian
boys and girls, who had been kept prisoners in the labyrinth, were
offered as prizes; and Theseus won their freedom for them, to the
general satisfaction of everyone, by defeating the unpopular Taurus
in a wrestling match.   It is not recorded how they won their
freedom in Demon's account, except that among the maidens who
were sent to Cnossus were two young men in disguise; and the
Deipnophoroi who took part in the Athenian festival of the Oschophoria were supposed to represent the mothers of the young people
sent to Crete, who brought them provisions for the voyage.

The version of Philochorus, though just as much a travesty of
the old-legend as that of Cleidemus, evidently claims to be the true
version from which the conventional story has arisen through
misunderstanding.   His tale of the adventures of Theseus when he
accompanied Peirithous on an attempt to carry off Persephone, no
longer queen of the underworld but daughter of Aedoneus, king of
the Molossians, is presented in the same way: his escape from deadly
peril, thanks to the intervention of Heracles, is supposed to have
been misunderstood as "a return from the house of Hades." [22]
Thus Theseus, the founder of so many religious cults at Athens,
who should be presented as a god-fearing hero if he is to win the

[20] Cf. Chap. 6, pp. 114–15 above.
[21] Cf. Chap. 4, p. 89 above.
[22] Fg. 46.

respect of Philochorus' contemporaries, does not engage in conflict with the gods of the underworld; nor does he offer violence to Minos, who, in this version, is doubtless a just judge rather than a cruel tyrant.   No doubt Philochorus is one of those writers mentioned by Plutarch, who regarded the story of the abduction of Helen as an unworthy libel on Theseus' character.   All the fragments of Philochorus which refer to Theseus represent him as a civilized negotiator, rather than a barbaric hero modelled after the style of Heracles.

The fragments of Philochorus are numerous enough to show that he treated other heroic figures in the same way, removing the barbaric elements in their character as well as the grotesque features of the legend.   He emphasized the soldierly qualities of Dionysus and defended him against charges of drunkenness and effeminacy.   He explained away the epithet διφυής ("of twin growth") applied to Cecrops by saying it referred to his exceptional tallness or his combination of Greek and Egyptian characteristics, and he insisted that Triptolemus travelled on a ship, not on a winged serpent.   His treatment of heroes together with his rejection of grotesque *aetia* and his attempts to clear the name of the Alcmaeonids from the charges of impiety have been discussed more fully in the previous chapter.[23]   The fragments enable us to form a fairly definite idea of the kind of rationalism that he favoured; it seems that he used rationalism, not as a weapon to discredit traditional religion, but rather in order to reinstate it and commend it to people who were not content to worship barbaric gods and heroes.

Evidently, then, when difficulty arose from contradictory tales about some legendary personage, Philochorus was ready either to explain away or to ignore any incidents which did not fit in with his characterization.   But this was not the only method available for solving the difficulty.   Hellanicus had explained the contradictory tales about Sardanapalus by maintaining that there were two kings of the same name, one an active conqueror, the other an indolent lover of luxury.   He had also been ready to duplicate characters in order to solve difficulties of genealogy; his use of this device in the case of Pelasgus, Ilus, and Oenomaus has already been discussed and there is no need to present the material again.[24]   It seems, indeed, to have been very generally believed that the tales about Minos did not all have reference to the same character; and

[23] Cf. p. 116 above.
[24] See Chap. 1 pp. 10–12 above.

in view of the long period of Cretan prosperity, modern critics have felt no difficulty in supposing that there was more than one King Minos. The Atthidographers, however, apparently did not accept this distinction between Minos the tyrant and Minos the lawgiver. There is no particular point in the versions of the tale of Theseus told by Cleidemus and Demon unless they wished to clear Minos from the charge of savage cruelty and make his character more consistent with that of a just lawgiver.

On the other hand, some of them certainly believed that there was more than one Eumolpus. Eumolpus was the founder of the Eleusinian mysteries, but he also appears in the guise of a Thracian king, with whom the Eleusinians joined in fighting against the Athenians in the reign of Erechtheus.[25] Naturally there was a difficulty in believing that the founder of the mysteries fought against Athens, and several versions which solved the difficulty are known. Thucydides tells how "the Eleusinians with Eumolpus fought against Erechtheus," but Isocrates represents Eumolpus as an invader from abroad who sought to dominate the whole of Greece, and says nothing of his connection with the Eleusinians.[26] It seems likely that this latter version was current in the fourth century, since Phanodemus, in whose account not one but two of Erechtheus' daughters sacrifice themselves to save their country, speaks of the threatening army as coming from Boeotia. The fragment of Philochorus unluckily speaks only in general terms of "the war which broke out, so that Eumolpus attacked Erechtheus." It is in Androtion's account that the distinction between the invader and the founder of the mysteries is most clearly set forth: the first Eumolpus, the invader, has a son Ceryx, whose son is called Eumolpus; then this second Eumolpus is father of the poet Musaeus, and it is Musaeus' son, Eumolpus III, who "started the mystic rites and became hierophant." In like manner Ister distinguishes Eumolpus the Thracian from the founder of the mystic cult, though the fragment does not show how he supposed them to be related to one another.[27]

---

[25] On this question see M. A. Schwartz, *Erechtheus et Theseus apud Euripidem et Atthidographos* 13–39.

[26] Th. 2.15.1; Isoc. *Paneg.* 68. Hellanic. F.40 does not show what version he preferred.

[27] Phanod. Fg. 3; Philoch. Fg. 14; Androt. Fg. 34 (cf. Chap. 4, p. 81 above); Ister Fg. 20, 21.

No attempt can be made here to explain the origin and development of the tale of Eumolpus,[28] and the fragments do not give us really adequate information about the way in which the different Atthidographers treated it.   It is, however, interesting to see that Androtion, who in various ways appears to agree with Hellanicus more closely than the others, builds up a complete sequence of five generations and distinguishes as many as three different characters bearing the same name.   In view of the different versions of the Theseus and Minos story it is unlikely that Androtion's explanation of this question was adopted by his successors.   Pausanias[29] remarks that "surely any one who is familiar with the old legends of the Athenians knows that it was Immaradus, the son of Eumolpus, who was killed by Erechtheus."   Since Cleidemus cleared Minos' character by making Theseus kill Deucalion, the son of Minos, after his father had already died, it is quite possible that Pausanias is referring to a similar attempt, made perhaps by an Atthidographer, to clear Eumolpus of blame and to clear Erechtheus of responsibility for his death by putting the two men in separate generations. The version which makes Eumolpus initiate Heracles into the mysteries[30] puts him even further back into the past.

The apparent diversity of views about the date of Eumolpus shows that there was room for disagreement about the chronology of Attic legendary history, despite the efforts of Hellanicus to reduce everything to order.   But the fragments tell us practically nothing of the individual views of the Atthidographers on chronology, and we cannot even be certain whether or not they accepted the succession of kings fixed by Hellanicus.   We learn that Philochorus followed Hellanicus in his date for Ogygus, and that he counted 189 years from Ogygus to Cecrops and made Cecrops reign 50 years.[31]   For the others, however, evidence of this kind is entirely lacking, and it is quite impossible to estimate what contribution they made to the study of the chronology of early Athenian history.   The fragments, for the most part, refer to their statements about the origins of religious cults, festivals, and temples, and give no indication whether or not they improved on the chronological

[28] For further references and discussion see Engelmann in Roscher's Lexicon *s.v.* "Eumolpos," Kern in *RE s.v.* "Eumolpos" (1).

[29] 1.27.4.

[30] Ister Fg. 20.

[31] Fg. 8, 10.   Cf. Chap. 6, p. 120 above.

scheme of Hellanicus.   A fragment of Ister shows that he mentioned
the trial of Cephalus in the reign of Erechtheus, but it does not
show whether he agreed with Hellanicus in placing an interval of
three generations between each of the four celebrated Areopagus
trials.[32]   The Parian Marble, however, agrees with Hellanicus in
putting the trial of Orestes over 300 years (nine generations) later
than that of Ares and Poseidon, and this is a fairly sure indication
that no Attic writer had been able to upset the scheme established
in the fifth century.   It is probable, therefore, that such alterations
and additions as were made after the time of Hellanicus were made
to fit in with this chronological outline.   If the portion of Aristotle's *Constitution of Athens* dealing with this early period were
preserved, a more definite conclusion might be possible.

Though it may be impossible to establish with any certainty how
permanently Hellanicus influenced the views of Attic historians
with regard to the chronology of very early times, there is no difficulty in showing how well he deserves in other ways to be considered
the founder of a tradition.   He should not be held responsible,
however, for the enduring interest of the Atthidographers in etymologies and *aetia*, because this was a legacy inherited from the old
Ionian historians and shared by the Atthidographers with many
other writers.   It is true that Herodotus and Hellanicus introduced
Ionian ἱστορίη to the Greek mainland, and for that reason they may
be held partially responsible for the permanent favour which these
features enjoyed among all historians who wrote at Athens.   But
we are concerned here with features peculiar to the *Atthides*,
rather than with characteristics which are to be found, in varying
degree, in the work of all Athenian historians.

Since the fragments not infrequently contain references to book
numbers, it is possible to show that different *Atthides* conformed to
a common model in their arrangement of material.   They all
started at the very beginning of Attic history.   There are fragments
which show that Hellanicus and Phanodemus discussed the question
of the origin of the Athenians and their claim to be autochthonous;
Philochorus spoke of the first ἄστυ founded by the Athenians when
they ceased to be homeless nomads; Androtion discussed the founding of Thebes by Cadmus, a digression which implies a comparison
with the legend of the founding of Athens.[33]

---

[32] Ister Fg. 19, Hellanic. F.169.   Cf. Chap. 1, pp. 15–17 above.
[33] Hellanic. F.161; Phanod. Fg. 7; Philoch. Fg. 4; Androt. Fg. 28–30.

Not only did they follow Hellanicus in starting at the very beginning (in contrast to Herodotus, who wanted to begin at the point when the conflict between Greeks and Asiatics first started), but they followed him also in allotting a substantial proportion of their work to the legendary period (herein differing markedly from Thucydides). Jacoby rejects Harpocration's reference to a fourth book of Hellanicus' *Atthis* [34] and thinks that it consisted of only two books, of which the first was devoted entirely to events of the regal period at Athens; but even if the *Atthis* did contain four books (which seems in itself not unlikely), one quarter of his work is a large proportion to set aside for the treatment of events in the remote past. There is a single reference, in Hesychius, to a twelfth book of the *Atthis* of Cleidemus; [35] but a more satisfactory fragment shows that in his third book he had not advanced beyond Cleisthenes, [36] so that he must have devoted two books to the legendary period and the obscure centuries after the Trojan Wars. There is not enough evidence to show how Phanodemus arranged his material nor in what books he wrote about Xerxes and Cimon. But the evidence for Androtion corresponds with that which is available for Cleidemus: the reference to a twelfth book [37] is probably incorrect, and other fragments show that he dealt with the Peisistratids in Book II; Book I is cited only for the founding of the Panathenaea by Erichthonius. For Philochorus the evidence is more complete. Suidas says that his *Atthis* contained seventeen books and Harpocration gives one reference to the sixteenth. [38] His first book was devoted to very early times, since the kings from Cecrops to Theseus were treated in his second book and Solon was not reached until the third. Ister's treatment of Attic mythology was evidently on quite a different scale and should not be brought into the comparison at all.

Since the later writers were in a position to apply detailed, annalistic treatment to a very much longer period than Hellanicus, it cannot be expected that they should devote so large a proportion of their work to mythical times as he did. Philochorus spent only four books in covering the period dealt with by Hellanicus; but he devoted half of this space to early times. Androtion reached the

[34] Note on Hellanic. F.44. Cf. Chap. 1, p. 14 above.
[35] Cleidem. Fg. 9.
[36] Fg. 8.
[37] Fg. 27. See Chap. 4, p. 79 above, note 18.
[38] Fg. 152.

end of the fifth century in his third or fourth book,[39] and, since he dealt with the tyrants in his second book, one whole book seems to have been taken up with the legendary and semi-legendary period. Thus, if we take into account only the period down to the end of the fifth century, we find that the difference between Hellanicus, Androtion, and Philochorus in their arrangement of material is not so great.

Again, in comparing the detailed annalistic treatment of events by the different Atthidographers, one should consider, not the proportion of a whole work devoted to this purpose, but the point at which an author begins to describe the events of each year under the name of its archon. The fragments of Philochorus seem to suggest that he first began to use an annalistic method some time after the middle of the fifth century and probably at the start of his fourth book.[40] The evidence for Hellanicus is unfortunately much less conclusive; one is almost entirely dependent on the statement of Thucydides that his treatment of the Pentecontaetia was too short and not detailed enough in its chronology.[41] But, if his *Atthis* contained four books—the same number which Philochorus devoted to the period covered by Hellanicus—it seems extremely probable that Hellanicus started his annalistic treatment of events in his fourth book and at the same point at which Philochorus began to describe events year by year. But no certain conclusion is possible here.

It is even more difficult to know how the different Atthidographers treated the period from the time of Theseus down to the point where they began to describe events year by year. The evidence of the fragments shows that all of them passed over this period comparatively quickly; and such few fragments as survive from this portion of their works usually refer to matters of antiquarian interest or such constitutional questions as would arise in the treatment of Solon and Cleisthenes. A reference to the fourth book of Demon's *Atthis* [42] shows that this author, of whom we know so little otherwise, was still dealing with the affairs of the later Athenian kings in that book, describing the coming of Melanthus, the Messenian, to Athens. This fragment, however, stands alone; no

[39] Fg. 10 and 11, dealing with the Thirty, are cited from Book III, Fg. 14, apparently referring to the battle of Arginusae, from Book IV.

[40] Cf. Chap. 6, pp. 121–23 above.

[41] Th. 1.97. Cf. Chap. 1, pp. 14–15 above.

[42] Demon Fg. 1.

other passage of narrative dealing with the period between the Trojan War and the time of Solon is quoted from any *Atthis;* [43] no *Atthis* is quoted even as authority for the conspiracy of Cylon or the code of Draco. The failure of the lexicographers to refer to them suggests that they found little information about this period in the *Atthides* which seemed useful for their purpose.[44] But people claimed in the fourth century to know more about this early period than is actually narrated by Herodotus and Thucydides, and, apart from the discussions in Aristotle, some of the material offered by Plutarch must go back to the fourth century. Plutarch, however, does not say so much about his sources in the *Solon* as in the *Theseus;* and conjecture must play a very large part in any attempt to reconstruct what the Atthidographers said about Draco or Cylon. A curious story about Draco, which Suidas [45] records without citing any authority, may perhaps be traced to some Atthidographer: that on a visit to Aegina Draco was greeted with great enthusiasm in the theatre and so many hats and articles of clothing were thrown at him that he was suffocated. Some *aetion* must be involved here, but Suidas has not explained the point of the story.

There are several fragments from the *Atthides* which refer to the reforms of Solon and Cleisthenes and to the Peisistratids, but we cannot establish with any certainty either the degree of detail in which they treated these topics or the method of approach adopted by the authors. Plutarch tells us that Androtion gave a heterodox interpretation of the Seisachtheia: that it was a monetary reform rather than a general cancellation of debts. It is also fairly clear that Androtion, whose work was known to Aristotle, had oligarchic sympathies; and, as an adherent of the "moderates," he can perhaps be held responsible for first presenting Solon as the ideal μέσος πολίτης.[46] These are valuable conclusions, as far as they go. But we cannot tell how far his version resembled or differed from the versions of others whose political affiliations are not known to us. It is quite certain that all of them had something to say about the

---

[43] Hellanic. (F.125) described the single combat between Melanthus and Xanthius, king of the Boeotians, but since the scholiast on Plato is citing Hellanicus primarily to show the descent of Codrus from Deucalion, Jacoby is uncertain whether the F. belongs to the *Deucalioneia* or the *Atthis.* The number of the book is not given.

[44] Harp. *s.v.* 'Απατούρια cites the second book of Ephorus for the tale of Melanthus.

[45] *S.v.* Δράκων.

[46] Cf. Chap. 4, pp. 83–84 above.

origins and development of Athenian democratic institutions; on this point the evidence of the lexicographers leaves no doubt. The difficulty is to know how complete their accounts claimed to be, how far they were consistent with one another, and to what extent they were accepted as accurate and authoritative by the Athenian public. Though Solon is frequently mentioned by the orators, it is not for the purpose of describing specific reforms of his, but in order to hold him up as an example of "the best of lawgivers" and the father of Athenian democracy.[47]

Very few fragments are available from *Atthides* dealing with the Persian Wars and the rise of the Athenian empire, and no detailed comparison is possible with the methods of the famous historians. An occasional anecdote is recorded, like that of the dog of Xanthippus or the "stratagem" of Themistocles in obtaining money to pay the sailors who were to man the Athenian ships in 480.[48] Other fragments relating to the early part of the fifth century show a tendency to patriotic exaggeration. According to Cleidemus only fifty-two Athenians fell at Plataea, all from the Aiantis tribe—a story apparently due to the fact that this tribe offered a special annual sacrifice in honour of the victory. Phanodemus exaggerated the glory of the Athenian victory of the Eurymedon by giving the barbarians the enormous number of six hundred ships. Other stories were told to illustrate the cleverness of the Athenian leaders: for example, how Cimon outwitted the enemy in Cyprus by giving orders for his death to be kept secret even from his own men.[49] The attempt of Philochorus to clear the Alcmaeonids of the various charges levelled against them is linked up with his effort to dispel any suspicion that the Delphic oracle was guilty of taking a bribe, and so with his general attitude in all matters affecting the traditional religion.[50]

The evidence does not permit us to compare the annalistic sections in the different *Atthides*, since it is only for Philochorus that an adequate collection of fragments is available. It has already been shown how some paragraphs in Thucydides approximate very closely to the style of Philochorus, and there are also occasions when the fourth century historians employed this bald

[47] I have discussed "Historical Allusions in the Attic Orators" in *CPh* 36 (1941) 209–29. For the references to Solon see especially pp. 221–24.

[48] Philoch. Fg. 84; Cleidem. Fg. 13.

[49] Cleidem. Fg. 14; Phanod. Fg. 17 and 18.

[50] Cf. Chap. 6, p. 116 above.

annalistic style. The close similarity between the two passages from Androtion and Philochorus quoted by Didymus [51] is also enlightening. Not only does it show that Philochorus used the same style as his predecessor, but it is some indication of the degree to which he made use of his work. If Philochorus gave an annalistic account of events covering a hundred and fifty years or more, as the fragments seem to indicate, it is only to be expected that in some of the lean years, when there was little to be recorded, he could do nothing except reproduce what earlier writers had said. The event which is recorded in such similar style by both Philochorus and Androtion, the settling of the boundaries of the sacred Orgas in the Megarid, took place in the year 350–49 B.C.; and no other events of any importance are known to have occurred in that year (though the next year is an extremely eventful one). There are, however, some other signs that he followed his predecessors closely on occasion. The scholiast on Aristophanes cites both Hellanicus and Philochorus for the minting of a gold coinage in the year of Antigenes,[52] implying that Philochorus does no more than follow Hellanicus. And from Harpocration's note it appears that he followed Androtion for the account of the revision of the citizen rolls in 346–5.[53]

There is, of course, nothing remarkable in the fact that Philochorus should follow the work of his predecessors nor does it reflect any discredit on him or justify any charge of plagiarism. It is interesting to note, however, that Clement of Alexandria, in his list of those who "stole material from Melesagoras," names Hellanicus, Androtion, and Philochorus rather than any other Atthidographers. It has already been pointed out that Clement is probably wrong in regarding Melesagoras as an early writer, previous to Hecataeus and Hellanicus; [54] but the passage has some bearing on the present discussion because it groups together just these three Atthidographers. The fragments suggest that Philochorus may have borrowed material from Androtion and Hellanicus and that the two later writers followed the methods of Hellanicus in the annalistic portion of their work; but no such evidence is available for Cleidemus, Phanodemus, or any of the others. Since, therefore, there is

[51] Cf. Chap. 6, pp. 128–29.

[52] Hellanic. F.172; Philoch. Fg. 120.

[53] Philoch. Fg. 133.

[54] Cf. Chap. 4, pp. 88–89 above. E. Stemplinger, *Das Plagiat in der griech. Literatur* 70–71, quotes this passage (*Strom.* 6.2.26) but inexcusably mistranslates it.

some evidence in the fragments to support the words of Clement in linking these three together apart from the other Atthidographers, it seems worth while to investigate more closely.

There is no reason to suppose that the Atthidographers referred to one another by name, until Ister, following the new Alexandrian custom, "collected the statements made by the different historians." [55] Aristotle, in his *Constitution of Athens*, is no more communicative about his literary obligations than Herodotus and Thucydides, and the custom of pretending to ignore the work of predecessors seems to have continued until the Alexandrian school brought in a change of fashion.   A fragment quoted by Harpocration shows that Philochorus mentioned Androtion by name in his discussion of the sacred utensils used in processions at Athens: "In former times the Athenians used the utensils bought out of the property of the Thirty, but later on Androtion provided others." [56] It seems likely that if Philochorus had made it clear that he was drawing on the *Atthis* of Androtion (as in fact he almost certainly is), Harpocration would have quoted enough to show this.   Again, Athenaeus at different times cites both Androtion and Philochorus for the old law at Athens forbidding the slaughter of a sheep before it had been shorn or had lambed, and then goes on to cite Philochorus for a time in Athens when the sacrifice of oxen was forbidden because the animals were becoming scarce. [57]   Here it looks as though Philochorus is not content to repeat what Androtion said but improves upon it, giving a further instance of a law forbidding certain sacrifices in the interests of the food supply.   Again, a confused scholion on the *Wasps* of Aristophanes does not make it quite clear what Philochorus had to say about the ostracism of Thucydides, son of Melesias; but here he could scarcely avoid drawing upon Androtion, who took special trouble to show that Theopompus was inaccurate and to distinguish the different people called Thucydides. [58] Philochorus' obligations to Hellanicus, on the other hand, are most clearly revealed in his statements about the chronology of very early Attic times; and Androtion's distinction of three different characters called Eumolpus is a good indication of how much he may have owed to Hellanicus, who duplicated mythical characters on several occasions.

---

[55] Fg. 11: τὰ παρὰ τοῖς συγγραφεῦσιν ἀναλεγόμενος.
[56] Philoch. Fg. 124.   Cf. Chap. 4, pp. 78–79 above.
[57] Philoch. Fg. 63, 64.
[58] Philoch. Fg. 95, Androt. Fg. 43, 44.

These scraps of evidence are not adequate to prove that Androtion and Philochorus were plagiarists or κλέπται, as Clement would call them, on a large scale. They are significant principally because evidence for annalistic treatment is available only for the *Atthides* of Hellanicus, Androtion, and Philochorus, not for the works of the other Atthidographers. From this point of view their work, so far as we know it, stands in a different category from that of the others. Lack of similar evidence for Cleidemus and Phanodemus may be no more than a coincidence; but the fact remains that, as an authority for details in Athenian history from the middle of the fifth century onward, Philochorus is cited far more frequently than his predecessors and as an annalistic account of events and a convenient book of reference his *Atthis* apparently superseded the earlier *Atthides*.

It follows, then, that in speaking of the general literary tradition of the Atthis and the common characteristics which all Atthidographers shared and inherited from one another, we must bear in mind that the evidence is uneven. Certain characteristics common to them all are admirably illustrated by the fragments: their concern with religious ritual and the mythological explanations of religious customs, with constitutional antiquities and the development of Athenian democratic institutions; their interest in the topography of Athens and Attica and the sacred associations of different Attic sites; and (though this point is less well illustrated) their interest in anecdote and biographical detail concerning both the political and literary figures of Attic history. Philochorus, as the latest in date of the Atthidographers properly so called, devoted a larger proportion of his work to annalistic treatment of historical events and accordingly commanded greater respect as an historical authority. And in his case the greater number of fragments available enables us to see the particular point of view which he took in attempting to give a consistent, rational, and credible account of the origins of traditional Attic religion.

INDEX

Acropolis, Athenian, 36f., 88, 107, 144.
Acusilaus of Argos, 88.
ἀδύνατοι, 119.
Aegeus, 12.
Aegina, 64, 159.
Aeschylus, 15f.
Aetia and aetiology, 8, 12f., 81, 96, 115, 118, 139, 142, 156, 159; *Aetia* of Callimachus, 60, 65, 72, 74, 139.
Alcibiades, 39, 53, 87, 95, 116.
Alcmaeonidae, 39, 116, 160.
Alcmeon, 31f.
Alexandria, 8, 36, 110, 112, 136–38, 144, 162; see also Callimachus, Lycophron, etc.
Amazons, 18f., 66, 150f.
Amelesagoras, see Melesagoras.
Amphictyon, Athenian king, 12, 17, 112, 118.
Amphictyonic league, 82.
Amphipolis, 33, 35, 40, 46, 79n., 80.
Andocides, 25f., 127, 134.
Andron of Halicarnassus, 81n., 87.
Androtion, 76–86, 117, 118, 128f., 141, 149, 154–59, 161–63.
Anecdote, 50f., 94f., 98f., 101f., 117.
Annalistic method, 14, 24f., 44–47, 51–53, 55, 85, 96, 121–34, 158, 160f., 163.
Antalcidas, 127f.
Anticleides, 59n.
Antigonus Gonatas, 107, 135.
Antiochus of Syracuse, 33, 146.
Antiquarian interest, 8, 36f., 63f., 69, 74, 96.
Apollodorus, *Bibliotheca*, 10–12, 16f.; *Chronica*, 5, 122.
Apollonius of Rhodes, 36n.; scholia on, 8n.
Archons, Athenian, 14, 20, 38, 41f., 85, 100, 103, 111, 112n., 121–25, 128–32; see also Annalistic method.
Areopagus, 13, 15–17, 67, 68, 75, 118f., 156.
Arginusae, battle of, 2, 5, 53, 79, 85, 103, 125.
Aristeides, 102, 103.
Aristogeiton, see Harmodius.
Aristophanes, 50, 91; scholia on, 2, 5, 14f., 24f., 85n., 87, 91, 105f., 121–26, 137, 144, 161, 162.
Aristotle, *Constitution of Athens*, 22, 23n., 49, 57, 59, 67f., 82–84, 92, 99–104, 119, 121, 125f., 156, 162.
Athenaeus, 49, 61, 63f., 74, 75, 90, 108f., 133, 137, 142.

Autochthony, claimed by Athenians, 15, 73, 139, 156.

Bias, national, 55f., 85, 95, 99; see also Exaggeration, Local patriotism, and National pride.
Bias, political, see Political sympathies.

Cadmus of Miletus, 1.
Callias, Peace of, 128.
Callimachus, 8n., 60, 65, 72, 74, 88, 136–39.
Cecrops, 12, 16f., 38, 111, 113, 115f., 118, 120, 139, 153, 155; daughters of, 88, 142.
Census of Attica, 115f., 120.
Chalceia, 70, 74f.
Charon of Lampsacus, 1, 3, 6, 88.
Choes, 70, 74.
Chronology; of early times, 9–12, 16f., 19f., 40, 42, 120, 155f.; of Greek history before Persian Wars, 40, 103; of 5th century, 1, 21, 24f., 40–44, 51–53, 96, 103, 121–26; of 4th century and later, 53, 96.
Cimon, 43, 50, 73, 95, 96, 99, 102, 126n., 160.
Citizen roll of Athens, 123, 161; see also Census of Attica.
Cleidemus, 57–69, 149–52, 154–57, 160; on religious matters, 61–64; on early times, 65–67, 149–52, 154f.
Cleisthenes, 14, 23, 39, 58, 67, 82, 111, 118n., 121, 159.
Cleitodemus, see Cleidemus.
Clement of Alexandria, 88f., 161–63.
Cleon, 84, 102, 124.
Cleophon, 102, 125.
Cnidus, battle of, 126–28.
Codrus, 19, 39, 90.
Colaenus, 13, 15, 70.
Colonization, 22, 33f., 40; see also Foundings of cities.
Colonus, 36, 47, 82, 117.
Conon, 126–28.
Constitution, Athenian, see Political institutions.
Conti, Natale, 105f.
Cranaus, 12, 16f.
Craterus, 91.
Cratippus, 56n.
Cults, see Religious discussion.
Cylon, 14, 37–39, 159.

164